Blockchain as a Technology for Environmental Sustainability

Editors
Maria José Sousa
ISCTE Instituto Universitário de Lisboa
Lisbon, Portugal
Tewabe Chekole Workneh
University of Verona
Verona, Veneto, Italy
Halvor Holtskog
Norwegian University of Science and Technology
Gjøvik, Norway

CRC Press is an imprint of the
Taylor & Francis Group, an **informa** business
A SCIENCE PUBLISHERS BOOK

Cover credit: Images provided by the first editor, Maria José Sousa. The images licensed under Envato Elements.

- Blockchain image: https://elements.envato.com/blockchain-39TSQUU (License code: B4ADZQMECJ)
- Sprout plant and bitcoin, growth of bitcoin crypto currency image: https://elements.envato.com/sprout-plant-and-bitcoin-growth-of-bitcoin-crypto--MH3NU5P (License code: RBV4FKLYTD)
- Blockchain technology concept image: https://elements.envato.com/blockchain-technology-concept-PEHGTVU (License code: ZK6BA42WL9)

First edition published 2024
by CRC Press
2385 NW Executive Center Drive, Suite 320, Boca Raton FL 33431

and by CRC Press
4 Park Square, Milton Park, Abingdon, Oxon, OX14 4RN

CRC Press is an imprint of Taylor & Francis Group, LLC

Library of Congress Cataloging-in-Publication Data (applied for)

ISBN: 978-1-032-19797-5 (hbk)
ISBN: 978-1-032-19799-9 (pbk)
ISBN: 978-1-003-26090-5 (ebk)

DOI: 10.1201/9781003260905

Typeset in Palatino Linotype
by Prime Publishing Services

Foreword

Blockchain Technologies are here and going mainstream. Decentralised Ledger Technologies are being implemented and adopted at an Enterprise and Government level, transforming existing processes, creating new services for employees and citizens and generating new monetisation channels that will undoubtedly create a new blockchain based value economy for the world. Latest Blockchain market research studies value a growth from $4.9 Billion USD to $67.4 Billion USD with a CAGR of 68.4% for the 2022–2026 period.* This is a testament to the growth of layer 1 blockchains and the industry focus to disrupt their processes and businesses with new technologies. Of course, emerging technologies need to align with core values of enterprises and governments and ESG goals are top priority.

According to the World Economic Forum's, Crypto Impact and Sustainability Accelerator (CISA) group, which I am part of, and which is focused on blockchain technologies and sustainability and how they can help achieve positive climate change impact, It is expected that, 3rd generation and future blockchain technologies will bring a dramatic reduction in energy consumption and environmental impact, as we move from proof of work to proof of stake and new consensus mechanism are introduced. Traditional bitcoin mining moves from hardware intensive compute to staking protocols. Above all, the application of blockchain in enterprises can be done in multiple ways, from on premise to cloud based deployments, making sure that industry standards are met when it comes to energy consumption and security. The layer 1 blockchains that are able to deliver on the overall value proposition that addresses new service creation or process improvement with a sustainability promise and delivery, are the ones that will win a long term spot in the market.

---

* https://www.marketsandmarkets.com/Market-Reports/blockchain-technology-market-90100890.html.

Alas, all is not good in the world of Web3. Blockchain technologies have challenges to overcome. Recent scandals like FTX, unfortunately drive a negative market outlook, so regaining trust in the overall ecosystem is important. Not only via stronger regulations, but also with safer and more easy-to-experience platforms for the user, that can drive adoption and implementation. The main objective of blockchain is still to deliver trust in an untrustworthy environment, tokenizing assets, registering transactions, being a ledger of trust to all, while at the same time, meeting environmental and ESG goals. Still, the economy needs to understand that blockchain is not crypto and that this technology is the cornerstone, the foundation of what web3 will be and what will be part of people's lives in the future.

Blockchain will be a driving force in the next decade. Having a clear understanding of what it is and how it will help improve the lives of people around the world whilst meeting businesses and country's ESG goals will be key for the reader to understand the dramatic transformation that we will see coming forth and new economies and value chains that will be created.

Joel Curado Silveirinha, MBA
Adjunct Professor, IE Business School, Madrid, Spain
Member, World Economic Forum's, Crypto Impact and Sustainability Accelerator (CISA) Group Managing Director
Casper Labs Enterprise AG, Zug, Switzerland

Preface

Blockchain technology is one of many emerging technologies that have the potential to solve some of the environmental problems we face today. Applications of blockchain to address environmental issues revolve around: monitoring and tracking supply chains, innovative financial instruments, peer-to-peer trading of tokenized assets, enabling decentralized energy systems, and common-pool resources. Applications of block chain technology could be useful in ensuring compliance with multilateral Environmental Agreements (MEAs) and progress on the Sustainable Development Goals (SDGs) implementation.

In this book the chapters are focused on the evolution of blockchain technology and its emergence as a promising tool to address various sustainability challenges across diverse industries. This book aims to explore the multifaceted applications of blockchain in promoting sustainability and driving innovations. Specifically, it focuses on its role in reducing electronic waste in healthcare, enhancing traceability and anti-counterfeiting measures, enabling information sharing for sustainable innovations, promoting environmental sustainability through smart contracts, waste management, and supply chains, as well as fostering green open innovation and sustainable forest supply chain management. Additionally, we will discuss blockchain's bibliometric review, its implications in the realm of cryptocurrencies, its potential in peer-to-peer carsharing, and its application in electronic voting systems.

The several chapter of the book will tackle the following specific topics:

Electronic Waste Reduction in Healthcare Finished Products Traceability and Anti-Counterfeiting: The healthcare industry generates significant amounts of electronic waste, posing environmental hazards and resource depletion. Blockchain technology offers a decentralized and transparent approach to track healthcare products' entire life cycle, ensuring proper disposal and recycling, reducing waste, and combating counterfeiting. By implementing blockchain-based traceability systems, healthcare providers can gain real-time insights into product origins, usage, and disposal, thereby fostering sustainable practices.

Blockchain as a Driver for Vertical and Horizontal Information Sharing for Sustainable Innovations: Blockchain's immutability and decentralization make it a powerful driver for vertical and horizontal information sharing. It enables organizations to collaborate effectively, share knowledge, and jointly work towards sustainability goals. By facilitating secure and transparent data sharing, blockchain promotes cross-sector innovation, helping various industries collectively tackle environmental and social challenges.

Blockchain Applied to Environmental Sustainability: In the Context of Smart Contracts: Integrating smart contracts with blockchain technology opens up new possibilities for environmental sustainability. Smart contracts automate and enforce sustainability-related agreements, ensuring compliance with eco-friendly practices. This essay explores the potential of blockchain-driven smart contracts in areas like renewable energy trading, carbon credit markets, and sustainable supply chain management.

Blockchain and Its Practical Implications in Waste Management Companies: A Practitioners' View: Waste management companies play a crucial role in achieving sustainable waste disposal and resource recovery. Blockchain can revolutionize waste management by improving transparency, accountability, and efficiency. This section investigates real-world case studies and applications of blockchain in waste management, highlighting its benefits from a practical perspective.

Improving Sustainable Supply Chains Using Blockchain Technology: A Review of the Literature: Supply chains are fundamental to sustainability efforts, and blockchain can enhance supply chain transparency, traceability, and accountability. This essay reviews existing literature on how blockchain technology can improve sustainable supply chain management, reduce environmental impacts, and promote responsible sourcing.

Blockchain x Green Open Innovation: Evidence and Trends for Sustainability: Green open innovation is essential for fostering sustainable advancements. Blockchain's decentralized nature enables trustful collaboration and knowledge sharing among innovators, researchers, and stakeholders. This section explores the trends, evidence, and potential of blockchain in driving green open innovation and accelerating sustainability-related projects.

Blockchain and a Sustainable Forest Supply Chain Management: A Literature Review: Forests are crucial ecosystems, and their sustainable management is vital for environmental conservation. This literature review investigates how blockchain can contribute to sustainable forest supply chain management by ensuring responsible sourcing, reducing illegal logging, and enhancing transparency in the timber industry.

Blockchain as a Technology for Environmental Sustainability: A Bibliometric Review: Bibliometric reviews offer insights into the growth and trends of research in a specific field. This section provides a bibliometric review of research related to blockchain technology's applications for environmental sustainability, highlighting the most influential works and identifying potential research gaps.

Cryptocurrencies - Advantages and Risks of Digital Money: Blockchain technology serves as the foundation for cryptocurrencies, revolutionizing the financial landscape. This section explores the advantages and risks of digital currencies, emphasizing their potential to reshape the financial sector while raising concerns about security, energy consumption, and regulatory challenges.

Blockchain-Based P2P Carsharing Towards Sustainability: Peer-to-peer carsharing can transform the transportation industry, reducing the number of vehicles on the road and promoting sustainable mobility. Blockchain's decentralized and secure nature enables efficient and transparent car sharing systems. This essay discusses how blockchain can facilitate peer-to-peer car sharing services, contributing to sustainable urban transportation.

A Proposal of a Blockchain-Based Electronic Election System: Blockchain's tamper-proof and transparent nature makes it an ideal candidate for enhancing the security and integrity of electronic voting systems. This section proposes a blockchain-based electronic election system, exploring the potential of this technology in ensuring fair and reliable voting processes.

The Relationship Between Blockchain Technology and Sustainability: A Bibliometric Analysis aims to explore the relationship between blockchain technology and sustainability through Bibliometric Analysis. A dataset of 313 articles published between 2017 and June 2022, containing the keywords 'blockchain' and 'sustainability,' was collected from the Web of Science database. The VOSviewer program was employed for Bibliometric Analysis. The analysis revealed four distinct research streams. The first stream examines how blockchain, in conjunction with Industry 4.0, can enhance sustainability in supply chains.

In conclusion, blockchain technology offers a vast array of applications that can significantly contribute to sustainability and drive innovations across various sectors. From reducing electronic waste in healthcare to fostering green open innovation and promoting sustainable forest supply chain management, blockchain's decentralized, transparent, and secure nature provides a robust foundation for building a more sustainable future. However, challenges such as scalability, interoperability, and regulatory concerns must be addressed to fully unlock the potential of blockchain for sustainability. With continued research, collaboration, and technological advancements, blockchain has the potential to revolutionize the way citizens and organisations address environmental and social challenges worldwide.

Maria José Sousa
ISCTE - Instituto Universitário de Lisboa
Tewabe Chekole Workneh
University of Verona
Halvor Holtskog
NTNU - Norwegian University of Science and Technology

Contents

CHAPTER 1

Electronic Waste Reduction in Healthcare; Finished Products Traceability and Anti-Counterfeiting

*António Pesqueira** and *Maria José Sousa*

1. Introduction

The healthcare industry is evolving more rapidly than ever. Most part of the new pharmaceutical products have shorter time to market, reduced from years to months, since the outbreak of the COVID-19 pandemic. Healthcare organizations can gain a competitive edge with the way the industry designs some of its processes and procedures to minimize waste and maximize resources, particularly when it comes to electronics, by redesigning these processes and procedures.

Several healthcare operations and functions, including supply chain and specialty care can be transformed by blockchain technology. Globally, blockchain's effectiveness has been recognized for its ability to reduce organizational costs, ensure regulatory compliance, and improve efficiency, by a variety of international organizations. Even though blockchain technology is still in the early stages of development

[1] Independent Researcher, Zug, Switzerland.
[2] ISCTE Instituto Universitário de Lisboa, Lisbon, Portugal.
Email: maria.jose.sousa@iscte-iul.pt
* Corresponding author: antonio.pesqueira@live.com

and adoption, healthcare stakeholders such as policymakers, regulators, industry, and organizational leaders, should be aware of its full spectrum of functionalities, as well as the legal, regulatory, and privacy concerns involved.

Blockchain is a relatively new technology in the healthcare industry. An example of how cryptographic chains can be used by the life sciences industry is demonstrated by some applications that are in the trial, and others that have already been adopted by organizations. Blockchain solutions can provide a new functionality in the pharmaceutical, biotechnology, and medical devices sector more focused on privacy, security and ethics, regarding sensitive data.

It is possible to utilize blockchain-based products' traceability and anti-counterfeiting applications, among others, to meet the needs of different modern healthcare processes which require a high degree of flexibility, methods, and processes to reduce electronic waste management, optimize resources, and create sustainable production and consumption practices (Kumar and Bhaskar, 2016). Blockchain-based technology differs from other technologies in that it does not seek to develop completely new products and services for the healthcare industry or others; rather, it reworks or improves existing processes in a way that unlocks efficiencies and new opportunities. We will explain in this paper how blockchain provides unique opportunities for providing resources efficiency throughout the supply chain process, from product traceability to logistics management to anti-counterfeiting measures.

Using techniques from mathematics and computer science, blockchain technology uses cryptography to encrypt every transaction into a unique digital signature and identifies the person who made the transaction. Therefore, in the following sections, we will provide a conceptual and practical understanding of blockchain applications, opportunities, and challenges with a focus on traceability and anti-counterfeit measures in the healthcare industry through a systematic literature review.

While analyzing healthcare applications, we will also understand through a systematic literature review all dependencies with major challenges, opportunities, use cases, future trends, and current knowledge. We will outline the major topics and concepts in the introduction and move on to a second section with details on the relevant topics in the literature review and all the related background and academic work.

Afterward, in Section 3, the methodology for the systematic literature review will be explained, as well as the process for planning and implementing it, and the results of the interviews will be discussed. The final section of this paper discusses the conclusions and recommendations for future work.

2. Literature Review

The objective of this paper is to present an overview of various real-world applications of blockchain technology, but it also aims at presenting a comprehensive literature review of the state-of-the-art and introduce key concepts used throughout the rest of the chapter.

National healthcare systems in all countries have a major challenge with a complex issue called the waste management of electrical and electronic equipment (Andrade et al., 2019), which includes all components, subassemblies, and consumables that are part of the equipment that is discarded. Many healthcare companies have changed their processes after the COVID-19 pandemic and millions of electronic devices are consumed and disposed of every year by the industry.

Keeping an eye on the environmental and health hazards posed by outdated information technology presents a challenge to healthcare organizations. The challenge of disposing of electronic waste (Gao, 2019) that is potentially harmful to human beings and has major environmental impact includes inflammable, combustible, ignitable, corrosive, toxic, reactive, injurious, and infectious waste (Schumacher and Agbemabiese, 2019). Our healthcare systems have been seriously affected by COVID-19, and the electronic waste management industry is no exception (Patibanda et al., 2020). Various organizations are investigating new methods or technology solutions, including blockchain.

It is not normally a priority in developing countries' healthcare systems to manage electronic waste, as international standards do not apply. In addition, there is an increased amount of potentially infected electronic waste that requires additional, careful handling (Zainu, 2020).

This research will highlight the major conclusions from the literature review that we have found to be valid for our research question: Can blockchain advancements in healthcare end-product traceability and anti-counterfeiting operations contribute to the reduction of healthcare electronic waste?

Some healthcare organizations in countries like Germany, the United States, and Switzerland have already published national recommendations on infectious waste management, but few guidelines specifically address COVID-19 waste management in most developing countries. In the healthcare waste stream, discarded computers, medical devices, healthcare electronics, and other consumer electronics like biometrics are the fastest-growing items due to their increased use and short life cycles. Although it seems counterintuitive that a product of technology can help solve the problem of electronic waste, the recycling issue can potentially be solved with blockchain.

Our analysis and search will show that blockchain can potentially be implemented to alleviate or even resolve the electronic waste concern using a variety of best practices like supply chain traceability and anti-counterfeiting measures. Organizations can use blockchain technology in recycling processes to trade old products and non-working components for digital tokens with full traceability of product codes, provenance, and distribution processes. As soon as blockchain technology is installed, data and information collected during the product distribution cycle are secure and tamper-proof.

We will, in the subsections that follow, have a detailed look at each of the relevant keywords present in this research, including blockchain, electronic waste management, finished healthcare products traceability, and counterfeit drugs.

2.1 *Blockchain*

Blockchain is a database that contains the history of different information pieces that allows the storage and validation of data transactions. Information is encoded as a string of blocks stacked one on top of another into a series called an immutable chain. Bitcoin, the first blockchain, establishes a log of transactions in bitcoin and stores the data, allowing for proof of ownership at any time. As a primary distinguishing characteristic of blockchain, their ledgers are publicly accessible, distributed, and mirrored globally on thousands of computers, called nodes, without central intervention or centralized control.

A blockchain is a decentralized database that enables data to be validated by decentralized communities, unlike centralized registers. In addition to the fact that bitcoin's blockchain is public, it is also trustworthy and secure, since its consensus mechanisms, how nodes verify new transactions, make clever use of subtle mathematics and computational brute force.

Because blockchain requires consensus, building and implementing applications to provide product traceability or conduct financial transactions, as well as distributing digital content, can be more challenging than relying on trusted intermediaries. The blockchain system is a peer-to-peer network where nodes share power and there is neither a central authority nor a single individual who can shut it down. If more than half of the participants attempt to overload a system, even if a central authority cripples one or more individuals or groups, the system will still function.

Furthermore, blockchain eliminates centralized power problems, enabling distributed and collective power because of the high capacity of end-to-end trace and trace on unit-level and the ability to ensure the validity of products to prevent sensitive disruptions (Agarwal et al., 2020) like counterfeit pharmaceutical supplies. To understand the entire logic of transparency, it is important to first understand attribution, which is the most important aspect of both asset ownership and contracts.

Blockchain is currently being extensively tested with different technologies including Radio Frequency Identification (RFID), which is an automated identification system based on radio frequency microwave transmission and is considered the natural evolution of traditional identification systems. With the continuous technological advances, the use of RFID systems with blockchain components has increased tremendously today, inspiring such devices to be one of the essential components of effective healthcare product quality monitoring and traceability, and part of smart communication, like the recent IoT (Internet of Things) applications.

Generally speaking, RFID technology includes a wide range of microdevices that can identify products and be used for quality measures, including pharmaceutical product identification, access control, medical device logistics monitoring, and centralized supply systems for sensitive products and services like vaccines or for gene therapies. When recording transactions in a blockchain, subsequent parties can see earlier transactions and who owns what. This offers several benefits, such as detecting counterfeit products, creating traceability, and validating serialization models.

Distributed ledgers allow everyone to see each transaction, instead of relying on a centralized database, thereby increasing transparency, and achieving higher quality levels, like in root cause analysis using multiple sensor measurements during shipment processes. Pharmaceutical manufacturers and medical device retailers can use the blockchain architecture to leverage the distributed, decentralized ledger to better share data within the healthcare supply chain, as well as to benefit from recycling, reuse, and remanufacturing in a fully circular economy to reduce costs and conserve the environment (Bekabil, 2020).

Blockchain technology could be used to create a network that allows us to share a great deal of information or metadata and provide us with more robust statistics on counterfeit prevalence and detection. Consequently, global supply chain security can be enhanced, and we may also be able to improve processes and frameworks for data management, because many markets cross over jurisdictions, and we may not be sharing data in the same way.

Aside from enhancing trust and transparency, the identification, tracking, and verification of drugs have many benefits. A blockchain application designed to support asset tracking in closed-loop supply chains, focusing on electronic waste management can be expected to provide full transparency, trust, and security, and to digitally streamline procedures (Khetriwal et al., 2009). One way blockchain can improve electronic waste management for healthcare is by better controlling and verifying medical devices, electronic hospital equipment, medical wearables, or trackers, or even computer devices used in healthcare or pharmaceuticals.

We can integrate blockchain architectures that are well established for tracking healthcare products from the manufacturer to the final disposal location (like hospitals), to overcome obstacles associated with copyright infringements, and to promote the circular economy of electronic waste (Forti et al., 2020).

2.2 *Electronic Waste Management in Healthcare*

Today, numerous factors determine whether supply chain management should change or adapt. The healthcare industry's attempt to achieve full supply chain traceability requires that organizations go through the product trail to ensure all actors have the data they need to contribute to the supply chain.

Specifically, the pharmaceutical supply chain begins with the production of raw materials, which are then produced into clinical treatment drugs, which are labeled and packed before entering the warehouse for distribution. During the final stages of a clinical trial or regulatory review, patients can get access to any investigational or previously approved medicine. There are several parties involved in this complex process. Companies today have no visibility into the entire supply chain of their products, and manufacturers and sponsors generally are unable to access the data being recorded at clinical sites in real-time. The result is that there is a continuous process of editing or adding of data by different parties, limiting data accuracy and consistency.

Additionally, manufacturers also spend a lot of resources ensuring compliance with regulatory requirements, since transparency and data visibility across the entire process is almost impossible. As a result, healthcare companies are unable to distinguish between different characteristics from patients, such as adverse outcomes or side effects.

In order to integrate the end-to-end traceability process, establishing a single source of truth across all participants is still one of the greatest challenges facing the industry. Future improvements will consist of linking all the process steps on a common layer, allowing all actors to access all relevant data points at every stage of the process and across the network, thereby reducing the effort and time required to find information. But very few companies are able to move forward with the first steps of creating more capabilities, increasing data ownership in healthcare product traceability, and facilitating reporting and analytics. A healthcare product's traceability involves many processes including the clinical supply chain process, medical devices, and healthcare support, among whom are contract manufacturing organizations and regulatory agencies.

In addition to quality assurance, cold chain management, and inventory management, manufacturers have to track and monitor all clinical shipments, from production through distribution, and connect them with medical providers and patients. There is no way for regulators or compliance companies to monitor real-time all supply chain activities, and it is impossible to define the product's lifecycle with a single platform.

Increasing process efficiency is achieved by digitizing manual or paper-based processes, which are tracked most of the time by paper documents that can involve many different repositories and are hard to reconcile. There are already several technology options being considered, especially blockchain for linking documents from those document repositories or digitizing the data contained in those documents so they can be analyzed by the many stakeholders while providing interoperable data points to make better decisions.

We will see that many people consider their prescription medications genuine, safe, and effective, yet counterfeit pharmaceutical products pose high-risk. Counterfeit medicines threaten the lives and health of millions of people around the world (Joon et al., 2017). It has been discovered that several pharmaceutical products have been duplicated, which include a variety of types like vaccines. Information exchange during the entire supply chain process should be fully trustworthy, transparent, traceable, and secure among all transactions and interactions with all involved stakeholders.

Manufacturers, public or private payers, wholesalers, and retailers are among the major stakeholders in a healthcare supply chain. Various healthcare organizations (HCOs) rely on these organizations for different parts of the drug distribution process, including manufacturing, transporting, and distributing the drugs. Using the supply chain, pharmacists can verify the authenticity of the drugs which are transported to their stores by tracking their movement in the supply chain right from production in the manufacturing plant.

Furthermore, spurious drugs put the healthcare industry at risk by creating revenue losses and exposing businesses to reputation loss and liability. Substandard medications have less therapeutic effect than the original and may not adhere to quality specifications. Spurious drugs can be any type of drug, prescription drugs, over-the-counter remedies, premium brand products, or very expensive prescription pills that are hard to obtain.

Further, they may also be illegal or prohibited substances, like benzodiazepines, which are regulated by law, or they may be based on market trends, i.e., based upon where they are popular. Weight loss products and antidepressants are related to several counterfeit drugs in the most developed countries. A less economically developed country (Forti, 2020) might have problems with antibiotics and anti-malaria medication or, worse still, counterfeit drugs like morphine analogs that are prone to misuse.

There are many risks associated with spurious drugs, but the three main risk factors are therapeutic failure, lack of safety, and the presence of unknown ingredients. Drugs that fail to manage or cure symptoms are therapeutic failures. Lab evaluation tests to identify the composition of drugs are the most common way to detect fraud in the healthcare industry. A sample of the drug is taken to a laboratory to determine the chemical composition. The true content of a product cannot be seen from a visual inspection because the product looks like the original, and most patients are unaware of this deception. Only in some cases spurious drugs deteriorate and present an inconsistent color owing to poor packaging or poor compression mechanisms.

Counterfeit goods have become more sophisticated though of poorer quality, so even experts cannot recognize them without laboratory testing. Counterfeit drugs have also become widespread in pharmacies as a result of increasingly complex global supply chains and advances in technology that make it easier for criminals to make and sell counterfeit drugs.

3. Conclusions

3.1 Limitations and Future Recommendations

A study was undertaken to understand different blockchain solutions for the management of electronic waste in healthcare contexts, and with particular attention given to product traceability and counterfeit. We presented results based on evidence, but with a recommendation for future large-scale empirical testing.

Ideally, it will be useful to design a broader scope that includes not only qualitative analysis but also quantitative research for better assessment of electronic waste management in the industry, particularly after the full impact of COVID-19.

As a result of the discussions, we were able to observe that novel ideas and solutions are springing up in smaller business units and even in large healthcare organizations, which are trying to incorporate an environmentally responsible logic for recycling and reusing materials.

A key conclusion of this chapter is that organizations' managers and decision-makers need to take a strategic approach to ensure that electrical and electronic waste is recycled efficiently.

3.2 *Final Considerations*

The research paper presents a comprehensive review of different studies that can provide creative solutions and new technology processes for the prevention of electrical and electronic equipment waste.

Since communication about healthcare products must take place in an easy, economically positive fashion through the entire production and laboring process, automatic identification systems are an essential part of the entire chain, and here blockchain due to its large applicability in core healthcare processes like product traceability and counterfeit prevention is an option (Li and Achal, 2020). We were able to better predict some of the results of the systematic literature review by starting with a study of some key concepts such as blockchain, electronic waste management, products traceability, and anti-counterfeiting models.

In the future, a cost-benefit analysis that focuses on different application cases of the different components of electronic waste management in complex healthcare systems is recommended, since the impacts of electronic waste materials recycling, and the potential for raw materials recovery, will contribute significantly to the bottom line. Future research objectives should include a large-scale case study, which will provide an opportunity for researchers to gain a deeper understanding of current development frameworks and use cases in the future, as described in the previous section.

Funding

This research is funded by the project "BLOCKCHAIN.PT (RE-C05-i01.01 – Agendas/Alianças Mobilizadoras para a Reindustrialização, Plano de Recuperação e Resiliência de Portugal na sua componente 5 – Capitalização e Inovação Empresarial e com o Regulamento do Sistema de Incentivos "Agendas para a Inovação Empresarial", aprovado pela Portaria N.º 43-A/2022 de 19 de janeiro de 2022).

References

Agarwal, S., Punn, N.S., Sonbhadra, S.K., Tanveer, M., Nagabhushan, P., Pandian, K.K. and Saxena, P. (2020). Unleashing the power of disruptive and emerging technologies amid COVID-19: A detailed review. arXiv preprint arXiv:2005.11507.

Andrade, D.F., Romanelli, J.P. and Pereira-Filho, E.R. (2019). Past and emerging topics related to electronic waste management: Top countries, trends and perspectives. Environ. Sci. Pollut. Res., 26: 17135–17151. doi: 10.1007/s11356-019-05089-y.

Bekabil, U.T. (2020). Industrialisation and environmental pollution in Africa: An Empirical Review. J. Resour. Dev. Manag., 69: 18–21.

Forti, V., Balde, C.P., Kuehr, R. and Bel, G. (2020). The Global E-Waste Monitor 2020: Quantities, Flows and the Circular Economy Potential. United Nations University; Bonn, Germany: United Nations Institute for Training and Research, International Telecommunication Union; Geneva, Switzerland: International Solid Waste Association; Rotterdam, The Netherlands.

Gao, Y., Ge, L., Shi, S., Sun, Y., Liu, M., Wang, B., Shang, Y., Wu, J. and Tian, J. (2019). Global trends and future prospects of E-waste research: A bibliometric analysis. Environ. Sci. Pollut. Res., 26: 17809–17820. doi: 10.1007/s11356-019-05071-8.

Joon, V., Shahrawat, R. and Kapahi, M. (2017). The emerging environmental and public health problem of electronic waste in India. J. Health Pollut., 7: 1–7. doi: 10.5696/2156-9614-7.15.1.

Khetriwal, D.S., Kraeuchi, P. and Widmer, R. (2009). Producer responsibility for E-waste management: Key issues for consideration—Learning from the Swiss experience. J. Environ. Manag., 9: 153–165. doi: 10.1016/j.jenvman.2007.08.019.

Kumar, B. and Bhaskar, K. (2016). Electronic waste and sustainability: Reflections on a rising global challenge. Mark. Glob. Dev. Rev., 1: 1–16. doi: 10.23860/MGDR-2016-01-01-05.

Li, W. and Achal, V. (2020). Environmental and health impacts due to E-waste disposal in China—A review. Sci. Total Environ., 139: 745. doi: 10.1016/j.scitotenv.2020.139745.

Patibanda, S., Bichinepally, S., Yadav, B.P., Bahukandi, K.D. and Sharma, M. (2020). Advances in Industrial Safety. Springer; Singapore: E-waste Management and its Current Practices in India, pp. 191–202.

Schumacher, K.A. and Agbemabiese, L. (2019). Towards comprehensive E-waste legislation in the United States: Design considerations based on quantitative and qualitative assessments. Resour. Conserv. Recycl., 149: 605–621. doi: 10.1016/j.resconrec.2019.06.033.

Zainu, Z.A. (2020). Development of policy and regulations for hazardous waste management in Malaysia. J. Sci. Technol. Innov. Policy, 5: 63–71.

CHAPTER 2

Blockchain as a Driver of Vertical and Horizontal Information Sharing for Sustainable Innovations

*Paul Kengfai Wan, Halvor Holtskog,**
Lizhen Huang and *Mariusz Nowostawski*

1. Introduction

Sustainable innovation has gained interest in the few past years as a lever to cope with international competitiveness (Boons et al., 2013). There are various explanations for the term 'sustainable innovation'; generally it is understood as the development of new products, processes, services and technologies that not only deliver improved economic performance, but also enhanced environmental and social performance, both in the short and long term (Bos-Brouwers, 2010; Tello and Yoon, 2008). As the world has grown increasingly services-oriented, sustainable innovation services have received much attention and are predicted to dominate the world economy eventually (Arnold et al., 2011).

A service is an intangible that is not directly involved with physical products (Wang et al., 2015). Over the last decade, there has been a swift expansion of service segments as they are huge profit drivers for firms (Cho et al., 2012). In the US, it is reported that over 90 percent of the

Norwegian University of Science and Technology.
* Corresponding author: halvor.holtskog@ntnu.no

GDP comes from the service industry (Wang et al., 2015). In order to stay competitive, firms gradually shift from tangible products to services. Information is the key in services. Based on Ackoff's DIKW pyramid (Data, Information, Knowledge and Wisdom), information is defined as data, put into context in such a way that they are usable and meaningful (*Ackoff, R.L. (1989). From Data to Wisdom. Journal of Applied Systems Analysis, 16: 3–9. - References - Scientific Research Publishing*, n.d.).

Information sharing is identified as one of the strongest elements in building trust in business-to-business relationships (Palmatier et al., 2006). In addition to establishing cooperation, information sharing can help a firm in making better informed decisions and enable quicker responses to market opportunities and changes (Cheung et al., 2011). There are different methods for facilitating information sharing (e.g., third-party intermediaries, proactive sharing based on the contractual agreement and verbal communication) (Wan et al., 2020). However, the level of reliability of distributed information using the above mentioned methods is often poor in quality and fragmented (Tian, 2017), because there is no verification of the information shared.

Blockchain technology has attracted interest as a digital tool for information sharing because of its immutable and distributed nature. The distributed nature of blockchain can also prevent single points of failure, a major weakness of a centralized system (Armbrust et al., 2010). In this chapter, we are going to explore two different innovative blockchain-based solutions to improve services from both patient- and human-centric aspects. The purpose of the chapter is to create knowledge regarding the extent to which blockchain can facilitate information sharing to enable sustainable innovative solutions To be more specific, this thesis analyses the vertical and horizon levels of information sharing based on the DIKW model.

2. Data, Information, Knowledge, and Wisdom (DIKW) Pyramid

Information is identified as an important element in a service chain for decision-making purposes. It is also important to define the term *information* as well as other terms like *data, knowledge,* and *wisdom (DIKW)*. DIKW has been discussed from the days of ancient Greek philosophers to more recent times, e.g., by Ackoff 1989 or Bellinger 2004. The hierarchy referred to variously as the 'Knowledge Hierarchy', the 'Information Hierarchy' or the 'Knowledge Pyramid' is fundamental here (*The Wisdom Hierarchy: Representations of the DIKW Hierarchy - Jennifer Rowley, 2007*, n.d.).

There are differences in the perspectives and properties of DIKW as presented by Ackoff (*From Data to Wisdom. Journal of Applied Systems Analysis, Ackoff, R.L. 1989, 16: 3–9. - References - Scientific Research Publishing,* n.d.), Rowley (*The Wisdom Hierarchy: Representations of the DIKW Hierarchy - Jennifer Rowley, 2007,* n.d.) and Bellinger (Bellinger, n.d.), but there are certain core elements of DIKW that are similar, nonetheless. They can be summarized as below, in Table 1. Ackoff's DIKW pyramid, as shown in Table 1, provides a graphical depiction of the hierarchy of data from the bottom layer to wisdom, the topmost layer. In this chapter, we focus on in information and data layers. Knowledge and wisdom are not discussed in this chapter.

Table 1. Terms and core elements of Ackoff's DIKW pyramid.

Term	Core elements	
Wisdom	the capacity to place knowledge into a framework and apply it to different situations	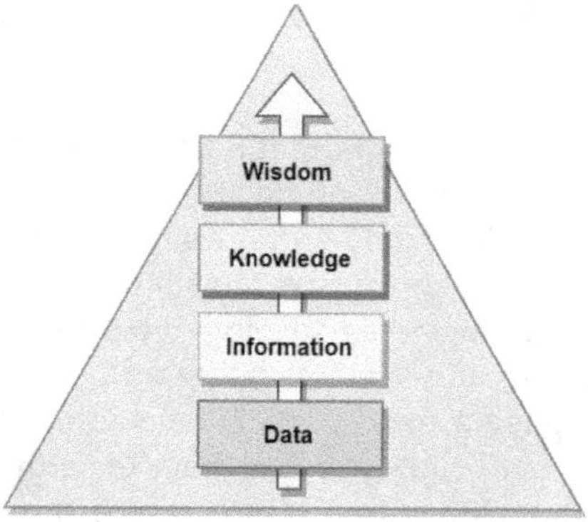 Wisdom Knowledge Information Data
Knowledge	information that explains the know-how about something that provides insights	
Information	data that is put in a context such that it is usable or meaningful	
Data	symbolic representation of objects, events and their environments	
		Ackoff (*Ackoff, R.L. (1989). From Data to Wisdom. Journal of Applied Systems Analysis, 16: 3–9. - References - Scientific Research Publishing,* n.d.)

Jennex (2017) revised the DIKW Pyramid, as shown in Figure 1, which includes decision-support technologies such as big data or IoT-based sensors that reflect the current state of the real world. With digital advances in technology such as the Internet of Things (IoT) and the explosion of big data, it has shifted from intuition-based decision-making to evidence-based decision-making (Chen et al., 2012; Madden, 2012; McAfee and Brynjolfsson, 2012).

This revised model shows that digital technologies can and must work together to improve the efficiency and effectiveness of decision-making. From both the revised and the traditional models, both information and data are needed to make decisions. In an earlier section, we saw that to

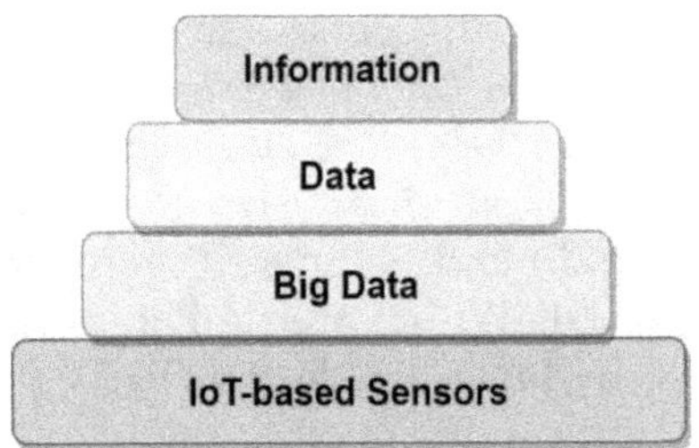

Figure 1. Revised DIKW Pyramid.

establish cooperation between two firms, information sharing is key. Furthermore, pure data will not create the meaning for decision-making without the context.

3. Blockchain Enables Vertical and Horizontal Information Sharing

Blockchain is a digital distributed computing network where no member in the network can falsify and control the information in the network. The blockchain data-storage structure makes tampering evident and this along with the consensus mechanism maintains the integrity of the data. This technology was developed because of central authorities having too much power and systems being vulnerable to abuses of power. Often having to engage a third-party intermediary is not desirable because it introduces intermediary service costs for storing and managing information (Wan et al., 2020). In addition, the firm must depend on and trust the services offered by the third party to facilitate information sharing. It is also not uncommon for the third-party intermediary to block access deliberately during disputes (Das et al., 2021).

Blockchain technology was developed to reduce some of the dependence on central authorities or to remove intermediaries and replace them with a distributed network in order to store, verify and safeguard the integrity of the transactions in multiple partnerships (Andoni et al., 2019). Unlike centralized authorities that have full access and greater control of data ownership, blockchain technology ensures all partners in the network hold their own copies of the ledger (Wan et al., 2020) or can access it in the open cloud (Andoni et al., 2019). Having access and control of data ownership, blockchain can enhance transparency and enable firms to make decisions with greater confidence.

The smart contract is another feature of blockchain that can facilitate information sharing without human intervention, thus reducing human error and cost, and it has caught the attention of researchers. Smart contracts are now used by researchers as an alternative way of governing and facilitating information exchange without the intervention of an intermediate third party (Filippi and Hassan, 2016; Nærland et al., 2017). Similarly, this has sparked an interest in research on how blockchain can play a role in information sharing both vertically and horizontally based on the DIKW model, as shown in Figure 1. In this way, the complexities of the service costs can be reduced by increasing efficiency, as with information sharing within a complex chain of actors (Deebak and AL-Turjman, 2021).

In this research work, information sharing is categorised into two types, namely vertical and horizontal information sharing which is based on the DIKW model. Horizontal sharing is information-information sharing where Firm A shares a piece of information with Firm B as shown in Figure 2. This type of horizontal information sharing shares the same element across two firms. While vertical sharing is a data-information sharing where Firm A shares data and Firm B receives a piece of information as shown in Figure 2. This vertical level information sharing is facilitated by blockchain layer which transforms data into information.

However, to the best of our knowledge, the only successful deployment of blockchain is still in cryptocurrencies like bitcoin. Nonetheless, there is a growing interest in integrating blockchain as part of digital transformation in industries such as healthcare (Chang and Chen, 2020; Khatri et al., 2021), smart cities (Li et al., 2019; Mora et al., 2021) and the energy sector (Erturk et al., 2019; Miglani et al., 2020). Therefore, in this chapter, we investigate how blockchain can improve information sharing in these three sectors, and present case studies.

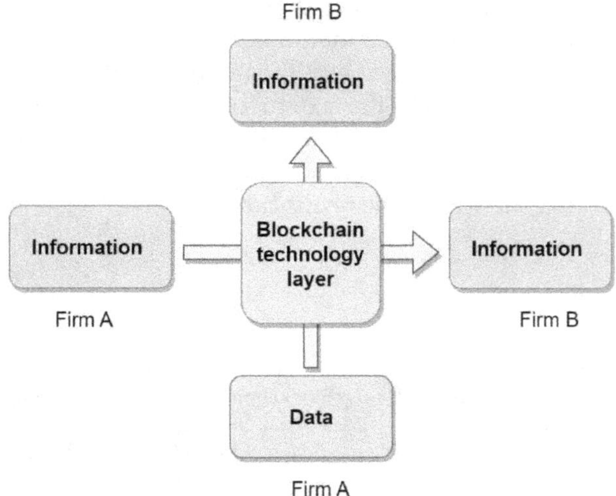

Figure 2. Vertical and horizontal information sharing.

4. Blockchain as a Driver for Sustainable Innovations through Information Sharing

Human health is one of United Nation's Sustainable Development Goals (UN SDG). We present two case studies of horizontal and vertical information sharing relevant to this goal. Both have been published in relevant journals: Journal of Medical Internet Research and Environmental Research. Here we give just a brief summary of the cases.

4.1 Case Study 1: Horizontal Information Sharing[1]

Purpose. Clinical decision support (CDS) is a tool that helps clinicians make decisions by generating clinical alerts to supplement their previous knowledge and experience. However, CDS generates a high volume of irrelevant alerts, resulting in alert fatigue among clinicians. This study aims to explore how a blockchain-based solution can reduce alert fatigue through collaborative alert sharing in the health sector, thus improving overall health care quality for both patients and clinicians.

Methods. We designed a four-step approach, as shown in Figure 3, to solve the problem of alert fatigue. First, we identified five potential challenges based on published literature through a scoping review. Second, a framework was designed to reduce alert fatigue by addressing the challenges identified using different digital components. Third, an evaluation was made by comparing MedAlert with other proposed solutions. We discuss these here, and end with the limitations of the research and possible future work.

Findings. Eight papers were selected and analysed out of 341 literature items collected initially. We identified five main key challenges: (1) data integrity, (2) privacy issues, (3) patient identity, (4) lack of secure information sharing and (5) the extent of patient's knowledge in the medical field, as shown in Figure 4. These challenges were taken into consideration in order to design and develop a feasible framework for reducing alert fatigue within the healthcare sector.

---

[1] **Reducing Alert Fatigue by Sharing Low-Level Alerts With Patients and Enhancing Collaborative Decision Making Using Blockchain Technology: Scoping Review and Proposed Framework (MedAlert).**
Wan, P.K., Satybaldy, A., Huang, L., Holtskog, H. and Nowostawski, M. Reducing Alert Fatigue by Sharing Low-Level Alerts With Patients and Enhancing Collaborative Decision Making Using Blockchain Technology: Scoping Review and Proposed Framework (MedAlert). J. Med. Internet Res. 2020; 22(10): e22013 doi: 10.2196/22013 (Wan et al., 2020).

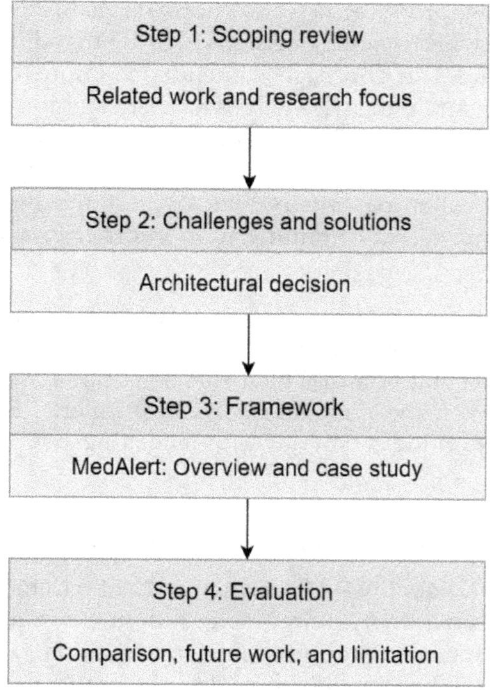

Figure 3. Research design process flow: 4-step approach. Figure from (Wan et al., 2020).

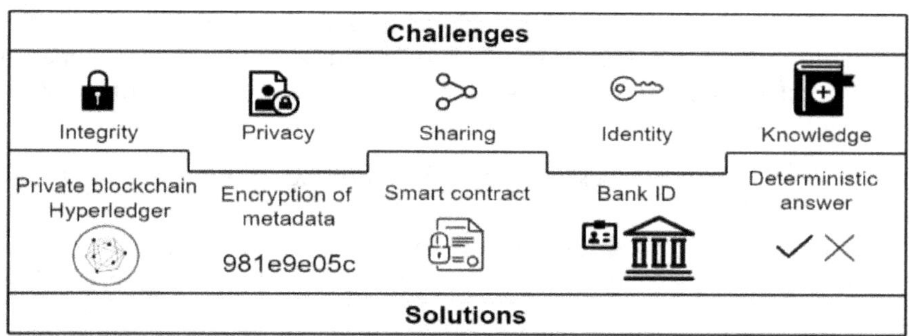

Figure 4. Architectural decisions in addressing the 5 key challenges. Figure from (Wan et al., 2020).

We found out that the better way of reducing alert fatigue is to get someone to attend to the alerts, rather than reducing the total number of alerts. This is because the benefit of reducing certain alerts remains unclear. Currently, there is a shift towards patient-centric data sharing to enhance collaborative decision-making, but relevant work remains limited due to the five challenges set out above.

In our work, we propose a new healthcare service by directing low-level alerts to patients when making decisions in order to reduce medical errors due to alert fatigue. This approach is intended to improve communication between clinician and patients. There is a risk of significant medical error when the clinician does not query and validate certain health information. The risk can increase when generated alerts are overridden due to alert fatigue. As such, when patients receive an alert and they are uncertain, they can enter into direct communication in order to avoid preventable errors.

Conclusion. We agree that the way to reduce alert fatigue is to get the attention of clinicians while they are attending to patients. However, there is no perfect solution in which clinicians can address all the alerts generated, not even the removal of irrelevant alerts. Blockchain-based technology can provide a new layer by engaging patients in providing responses.

4.2 *Case Study 2: Vertical Information Sharing*[2]

Purpose. Indoor air quality (IAQ) is an important parameter in protecting the occupants of an indoor environment. Indoor environments with poor ventilation have increased airborne virus transmission. Ultimately, indoor environments will become hotspots for the transmission of airborne viruses. Infection risk assessments can estimate virus transmission via airborne routes. From our literature search, we did not identify any systems integrating risk assessments with smart sensors to support experts in indoor environments in their decision-making. One of the reasons for this is that the complex set of stakeholders involved make information sharing difficult.

Methods. AIRa is a blockchain-based prototype which integrates CO_2 sensor data with infection risk assessments from a post-pandemic perspective. Our novel blockchain-based alert framework integrates infection risk assessment tools and IoT to enable early decision-making, as shown in Figure 5.

---

[2] **Automated infection risks assessments (AIRa) for decision-making using a blockchain-based alert system: A case study in a representative building.**
Paul Kengfai Wan, Lizhen Huang, Zhichen Lai, Xiufeng Liu, Mariusz Nowostawski, Halvor Holtskog and *Yongping Liu. Automated infection risks assessments (AIRa) for decision-making using a blockchain-based alert system: A case study in a representative building. Environmental Research, https://doi.org/10.1016/j.envres.2022.114663* (Wan et al., 2023).

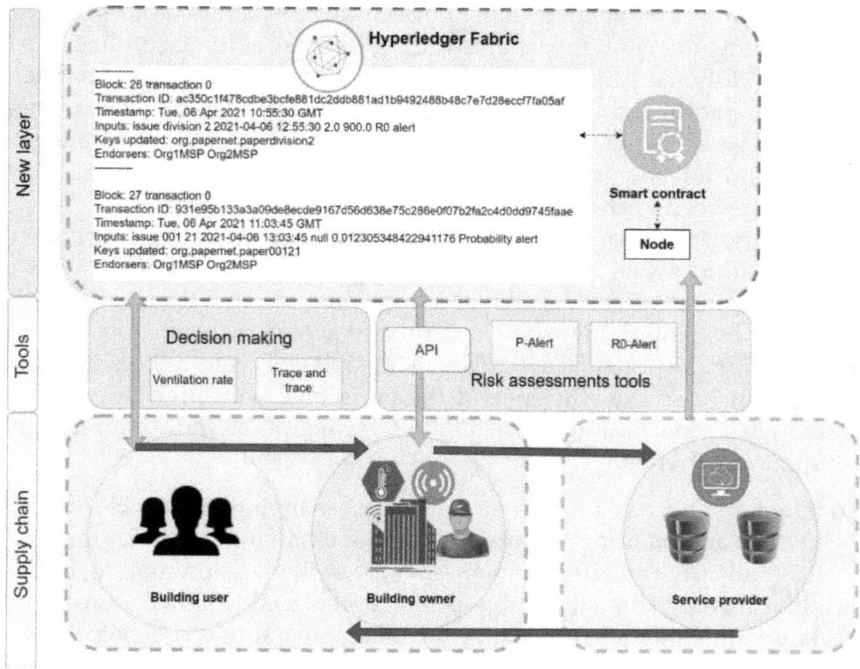

Figure 5. Blockchain-based alert system framework (AIRa). Figure from (Wan et al., 2023).

Multiple stakeholders are involved in requesting data to obtain information. The workflow of requesting building sensor data and information runs from a building user (requestor) to the building owner and finally to a third-party service provider. It is usual for the building owner to engage a third party to store and manage sensor data, which sometimes requires a service fee. Ultimately, this reduces the efficiency with which data can be obtained and makes collaborative work and analysis challenging.

In this framework, we used two infection risk assessment models: (1) the probability of infection (P), and (2) the basic reproductive number (R0). The experiments were conducted in an office on a floor with twelve meeting rooms with doors, and open desks with seating. No HVAC system is installed on the office floor, whose occupants rely instead mostly on natural ventilation. A sensor for monitoring indoor air quality such as CO_2 levels is installed in the meeting room The area of the room is $12m^2$ with 3 m height.

Findings. We found that with information sharing, this solution can be used as a long-term approach in protecting occupants from viral infections in an indoor environment by sending an alert to building users, thus enabling infection risk analysis using internet of things, unlike other alerting solutions. Although our framework can only cover a confined space like an office, it can play an important role and could be used long-term in assessing the condition of indoor environments using built-in sensors and infection risk assessment tools, as well as passive tracking without the need to check-in.

We also found out that blockchain can increase the transparency of data within a vast amount of data in a centralized database. With a timestamp alert that highlights anomalous events, transparency is enhanced by pinpointing the exact time at which to begin tracing and investigating the potential spread by using datasets around the timepoint. This can increase the efficiency of digital forensics through deeper and more holistic analyses by using all the information on indoor air quality around that timepoint.

Conclusion. Our work opens up new services by integrating blockchain technology and building sensors with risk assessments to generate an alert of potential airborne virus transmission for early decision-making by the building owner. This immutable event can enhance digital forensics by tracking other anomalies that may have occurred at that timepoint. AIRa can be useful in refining both public and risk management strategies, thus enabling epidemiologists to reduce the risk of indoor infection. However, it is still at an early stage, and a lot of work still needs to be done to realize this model as a new service in making smart buildings smarter.

5. Discussion

Case study 1 explored how blockchain-based solution can enhance information sharing on horizontal level in the healthcare sector. In this case study, the main issue that was identified is that information does not flow beyond a clinician's own health institution owing to strict regulations over sharing of information with other health institutions. Currently, clinicians receive a high number of low priority alerts and are gradually becoming less responsive to them, which opens door to preventable medical errors (Rayo et al., 2015).

The blockchain-based framework (MedAlert) in **case study 1** can create a channel for horizontal information alert flow from clinicians to patients. When a low priority alert is generated from the CDS system, this triggers the smart contract embedded in MedAlert and shares the alert with the patients. All the events are stored in blockchain where patients can also view the events. MedAlert can act as common layer of information sharing to capture the attention of either patients or clinicians, so they attend to the alert and thus prevent a potential medical error. This novel solution can move toward a clinician-patient collaborative decision-making process to avoid potential medical errors resulting from alerts being ignored. Ultimately, it can improve the quality of the health care domain by increasing better patient outcomes and reducing physician burnout.

Case study 2 explored a smart building by investigating blockchain technology for vertical data-information sharing. Most of the new buildings with embedded sensors focus on the indoor air quality (IAQ) parameters to protect occupants in the indoor environment. Most of the sensor data are stored and managed by third-party service-providers. Although data sharing of IAQ is less restricted than healthcare because it does not involve PII data, often it may come with a service fee from the service provider. This reduces the overall efficiency to obtain data for analysis and improvement work.

Case study 2, a blockchain-based framework (AIRa), aims to assess the potential viral infection based on CO_2 concentration level in the room and send an alert to building owners to take necessary actions to protect indoor occupants. The CO_2 concentration level in the room is continuously measured using a sensor and used as input to estimate the risk. When the CO_2 concentration level exceeds the pre-set R0 risk threshold, it triggers the smart contract embedded blockchain to send an alert to the building owner. The building owner can either take actions such as increasing the ventilation rate or simply ignore it, depending on the alert.

As mentioned in Section 2, information is data placed in context: sharing pure data will not create any meaning or aid decision-making. Per this vertical information sharing explored in **case study 2,** a new service can be created for decision-making by transforming IAQ data to a new parameter to be considered in a smart building to reduce the risk of indoor infection. Ultimately, this can push towards creating a smarter building.

Information sharing can not only establish new collaborations, but also open up new services. However, when it comes to information sharing, one key concern is trustworthiness of data. Blockchain technology offers better solutions, which can be a complementary tool in ensuring that shared information is not tampered with. This is very important to ensure that good decisions can be developed from correct information in all sectors. **Both case studies** are examples where a high level of trust is crucial when it comes to making decisions.

6. Limitation, Challenges, and Future Work

Information sharing can create new services to enable more collaborative decision-making. Challenges such as ethical and privacy issues could be significant, as highlighted in **case study 1.** It is difficult to find the right balance of information sharing with new partners, despite the potential improvements in the effectiveness of decision-making. Although this technology can enable new services and improve effectiveness, future work on firms' perceptions and acceptance of such newly enabled services is important to ensure smooth deployment.

In order to implement **case study 2** in a real-world scenario, future work in addressing other key requirements. Furthermore, while a quantitative evaluation can complement the qualitative evaluation, it can also demonstrate the actual performance evaluation, which is currently lacking in the research domain. Without documented performance in real industrial situations, it is often difficult to persuade enterprises and gain support from the top management to incorporate this technology as part of digital transformation.

General Data Protection Regulations (GDPR) with relevance to blockchain applications in the healthcare sector should be examined. GDPR regulates the collection, processing and securing of personal data to minimize the risks of privacy violation. However, GDPR could be violated due to the immutable nature of blockchain. Although in **case study 1,** the framework only stores metadata, not actual health data, and further ensures higher levels of privacy protection, researchers should work on aspects of the framework. Art 17 (the right to erasure or to be forgotten) in cases when users want to have their data completely erased or deleted, is one area to focus on.

7. Conclusion

The potential roles of blockchain to enable innovative, sustainable solutions through information are summarized below:

- Blockchain enables new forms of vertical information (data-information sharing). It is more usual to share data or information on the horizonally based on the DIKW model. Embedding smart contracts in blockchain enables autonomous execution of processes once a set of predefined rules is met. With this, data can easily be placed in their meaningful context during the sharing process, as demonstrated in case study 2. AIRa can lead to new ways of sharing data and information and ultimately increase efficiency to take appropriate action within a building.

- Another outcome of this research on information sharing is creation of new and innovative services with the help of digital tools like blockchain. Some papers argue that information sharing can often result in conflicts of interest. However, based on case study 1, MedAlert has benefits such as reducing preventable medical error through collaborative care and clinician burnout. However, health related information sharing is often difficult due to strict health regulation.
- In reality, data and information layers, based on Ackoffs's DIKW pyramid are fragmented and disjointed making transformation of data to meaningful information challenging. This may be a result of information silos and asymmetry which may be due to poor data sharing practices. As explained in this work, blockchain can play a role as a bridge in sharing information through vertical and horizontal modes with a more human centric approach as shown in both case 1 and 2.

References

Ackoff, R.L. (1989). From data to Wisdom. Journal of Applied Systems Analysis 16: 3–9. - References—Scientific Research Publishing (n.d.). Retrieved 8 February 2023, from https://www.scirp.org/(S(lz5mqp453ed%20snp55rrgjct55))/reference/referencespapers.aspx?referenceid=2918522.

Andoni, M., Robu, V., Flynn, D., Abram, S., Geach, D., Jenkins, D., McCallum, P. and Peacock, A. (2019). Blockchain technology in the energy sector: A systematic review of challenges and opportunities. Renewable and Sustainable Energy Reviews, 100: 143–174. https://doi.org/10.1016/j.rser.2018.10.014.

Armbrust, M., Fox, A., Griffith, R., Joseph, A.D., Katz, R., Konwinski, A., Lee, G., Patterson, D., Rabkin, A., Stoica, I. and Zaharia, M. (2010). A view of cloud computing. Communications of the ACM, 53(4): 50–58. https://doi.org/10.1145/1721654.1721672.

Arnold, J.M., Javorcik, B.S. and Mattoo, A. (2011). Does services liberalization benefit manufacturing firms?: Evidence from the Czech Republic. Journal of International Economics, 85(1): 136–146. https://doi.org/10.1016/j.jinteco.2011.05.002.

Bellinger, G. (2004). Knowledge Management. Retrieved 8 February 2023, from http://www.systems-thinking.org/kmgmt/kmgmt.htm.

Boons, F., Montalvo, C., Quist, J. and Wagner, M. (2013). Sustainable innovation, business models and economic performance: An overview. Journal of Cleaner Production, 45: 1–8. https://doi.org/10.1016/j.jclepro.2012.08.013.

Bos-Brouwers, H.E.J. (2010). Corporate sustainability and innovation in SMEs: Evidence of themes and activities in practice. Business Strategy and the Environment, 19(7): 417–435. https://doi.org/10.1002/bse.652.

Chang, S.E. and Chen, Y. (2020). Blockchain in health care innovation: Literature review and case study from a business ecosystem perspective. Journal of Medical Internet Research, 22(8): e19480. https://doi.org/10.2196/19480.

Chen, H., Chiang, R.H.L. and Storey, V.C. (2012). Business intelligence and analytics: From big data to big impact. MIS Quarterly, 36(4): 1165–1188. https://doi.org/10.2307/41703503.

Cheung, M.-S., Myers, M.B. and Mentzer, J.T. (2011). The value of relational learning in global buyer-supplier exchanges: A dyadic perspective and test of the pie-sharing premise. Strategic Management Journal, 32(10): 1061–1082. https://doi.org/10.1002/smj.926.

Cho, D.W., Lee, Y.H., Ahn, S.H. and Hwang, M.K. (2012). A framework for measuring the performance of service supply chain management. Computers & Industrial Engineering, 62(3): 801–818. https://doi.org/10.1016/j.cie.2011.11.014.

Das, M., Tao, X. and Cheng, J.C.P. (2021). A secure and distributed construction document management system using blockchain. pp. 850–862. *In*: Toledo Santos, E. and Scheer, S. (eds.). Proceedings of the 18th International Conference on Computing in Civil and Building Engineering. Springer International Publishing. https://doi.org/10.1007/978-3-030-51295-8_59.

Deebak, B.D. and AL-Turjman, F. (2021). Privacy-preserving in smart contracts using blockchain and artificial intelligence for cyber risk measurements. Journal of Information Security and Applications, 58: 102749. https://doi.org/10.1016/j.jisa.2021.102749.

Erturk, E., Lopez, D. and Yu, W.Y. (2019). Benefits and risks of using blockchain in smart energy: A literature review. Contemporary Management Research, 15(3): Article 3. https://doi.org/10.7903/cmr.19650.

Filippi, P.D. and Hassan, S. (2016). Blockchain technology as a regulatory technology: From code is law to law is code. First Monday. https://doi.org/10.5210/fm.v21i12.7113.

Jennex, M.E. (2017). Big Data, the Internet of Things, and the revised knowledge Pyramid. ACM SIGMIS Database: The DATABASE for Advances in Information Systems, 48(4): 69–79. https://doi.org/10.1145/3158421.3158427.

Khatri, S., Alzahrani, F.A., Ansari, M.T.J., Agrawal, A., Kumar, R. and Khan, R.A. (2021). A systematic analysis on blockchain integration with healthcare domain: Scope and challenges. IEEE Access, 9: 84666–84687. https://doi.org/10.1109/ACCESS.2021.3087608.

Li, J., Greenwood, D. and Kassem, M. (2019). Blockchain in the built environment and construction industry: A systematic review, conceptual models and practical use cases. Automation in Construction, 102: 288–307. https://doi.org/10.1016/j.autcon.2019.02.005.

Madden, S. (2012). From databases to big data. IEEE Internet Computing, 16(3): 4–6. https://doi.org/10.1109/MIC.2012.50.

McAfee, A. and Brynjolfsson, E. (2012, October 1). Big Data: The Management Revolution. Harvard Business Review. https://hbr.org/2012/10/big-data-the-management-revolution.

Miglani, A., Kumar, N., Chamola, V. and Zeadally, S. (2020). Blockchain for Internet of Energy management: Review, solutions, and challenges. Computer Communications, 151: 395–418. https://doi.org/10.1016/j.comcom.2020.01.014.

Mora, H., Mendoza-Tello, J.C., Varela-Guzmán, E.G. and Szymanski, J. (2021). Blockchain technologies to address smart city and society challenges. Computers in Human Behavior, 122: 106854. https://doi.org/10.1016/j.chb.2021.106854.

Nærland, K., Müller-Bloch, C., Beck, R. and Palmund, S. (2017). Blockchain to rule the Waves—Nascent design principles for reducing risk and uncertainty in decentralized environments. International Conference on Interaction Sciences. https://www.semanticscholar.org/paper/Blockchain-to-Rule-the-Waves-Nascent-Design-for-and-N%C3%A6rland-M%C3%BCller-Bloch/3f73fd41e1ccd35c03bf88ffd5a7334d88d20ada.

Palmatier, R.W., Dant, R.P., Grewal, D. and Evans, K.R. (2006). Factors influencing the effectiveness of relationship marketing: A meta-analysis. Journal of Marketing, 70(4): 136–153. https://doi.org/10.1509/jmkg.70.4.136.

Rayo, M.F., Kowalczyk, N., Liston, B.W., Sanders, E.B.-N., White, S. and Patterson, E.S. (2015). Comparing the effectiveness of alerts and dynamically annotated visualizations (DAVs) in improving clinical decision making. Human Factors, 57(6): 1002–1014. https://doi.org/10.1177/0018720815585666.

Tello, S. and Yoon, E. (2008). Examining drivers of sustainable innovation. Journal of International Business Strategy, 8(3): 164–169.

The wisdom hierarchy: Representations of the DIKW hierarchy—Jennifer Rowley (2007) (n.d.). Retrieved 8 February 2023, from https://journals.sagepub.com/doi/10.1177/0165551506070706.

Tian, F. (2017). A supply chain traceability system for food safety based on HACCP, blockchain & Internet of things. 2017 International Conference on Service Systems and Service Management, 1–6. https://doi.org/10.1109/ICSSSM.2017.7996119.

Wan, P.K., Huang, L. and Holtskog, H. (2020). Blockchain-enabled information sharing within a supply chain: A systematic literature review. IEEE Access, 8: 49645–49656. https://doi.org/10.1109/ACCESS.2020.2980142.

Wan, P.K., Huang, L., Lai, Z., Liu, X., Nowostawski, M., Holtskog, H. and Liu, Y. (2023). Automated infection risks assessments (AIRa) for decision-making using a blockchain-based alert system: A case study in a representative building. Environmental Research, 216: 114663. https://doi.org/10.1016/j.envres.2022.114663.

Wan, P.K., Satybaldy, A., Huang, L., Holtskog, H. and Nowostawski, M. (2020). Reducing alert fatigue by sharing low-level alerts with patients and enhancing collaborative decision making using blockchain technology: Scoping review and proposed framework (MedAlert). J. Med. Internet Res., 22(10): e22013. https://doi.org/10.2196/22013.

Wang, Y., Wallace, S.W., Shen, B. and Choi, T.-M. (2015). Service supply chain management: A review of operational models. European Journal of Operational Research, 247(3): 685–698. https://doi.org/10.1016/j.ejor.2015.05.053.

CHAPTER 3

Blockchain Applied to Environmental Sustainability

In the Context of Smart Contracts

Tewabe Chekole Workneh

1. Introduction

In an era marked by growing environmental concerns and an urgent need for sustainable solutions, the intersection of blockchain technology and environmental sustainability is emerging as a powerful force for positive change. Blockchain, originally developed as the enabling technology for cryptocurrencies like Bitcoin, has evolved far beyond its original application. It now provides an innovative and robust framework for addressing some of the most pressing challenges facing our planet. In the context of smart contracts, the potential of blockchain to revolutionize the way we approach environmental sustainability becomes even more apparent. The environmental challenges of our time, including climate change, resource scarcity, and ecosystem degradation, require innovative solutions that transcend geographic and political boundaries. Blockchain technology, with its decentralized and transparent nature, has the ability to not only track and verify environmental efforts, but also incentivize positive behaviors. This revolution in accountability and transparency, combined with the automation capabilities of smart contracts, presents a unique opportunity to revolutionize the way we engage in environmental sustainability.

Admas University, Addis Ababa, Ethiopia.
Email: tewabechekole.workneh@univr.it

2. Blockchain and Environmental Sustainability

In 2008, Satoshi Nakamoto made the first public disclosure relating to blockchain. Peer-to-peer networks and encryption became operational. Private Key Cryptography, Peer to Peer Networks, and Programs are the three essential elements of a blockchain (the blockchain protocol). Cryptocurrencies represent the primary use of blockchain technology. A peer-to-peer network and distributed timestamping server are used to manage the database of the blockchain. Blockchain platforms have enabled the provision of more services as smart contract technologies have progressed over time.

Blockchain is a technique of storing data that makes it difficult or impossible to alter, hack, or cheat the system. A blockchain maintains a digital log of transactions that is duplicated and distributed across the blockchain's network of computer systems. As the systems exchange data, the data recorded in each system changes, which anyone can readily trace and find out. It is now suggested for a variety of professions, including finance and the military, because of its features.

A blockchain creates a digital transaction ledger that is kept secure by a computer network and is impossible to hack or alter. Even if it has been hacked, finding the information is tough. Individuals can trade directly with each other without the need for an intermediary such as a government, bank, or other actors, thanks to technology. One of the benefits of blockchain technology is that it eliminates the need for intermediary entities, resulting in less data loss and changes in character or composition.

The basic purpose of blockchain technology is to allow for the storage and distribution of digital data without the ability to alter it. A blockchain, in this context, is a basis for immutable transaction records that cannot be changed, erased, or destroyed. A distributed database shared by nodes on computer networks is known as a blockchain. Like a database, a blockchain stores information digitally and electronically. Blockchains are well known for their involvement in cryptocurrency systems like bitcoin and other financial transactions, as they provide a secure and decentralized record of transactions. The concept of a blockchain ensures the accuracy with which something is replicated, as well as the security of a data record and trust, without the need for a third party.

The layout of data is a key distinction between a standard database and a blockchain. Information is organized in a blockchain into blocks, each comprising a set of data. When a block is full, it is closed and linked to the preceding block, forming a data chain known as a blockchain. Any additional data that comes after this freshly added block is merged into a new block, which is then added to the chain once it is complete. A database, on the other hand, organizes data into tables, but a blockchain, as

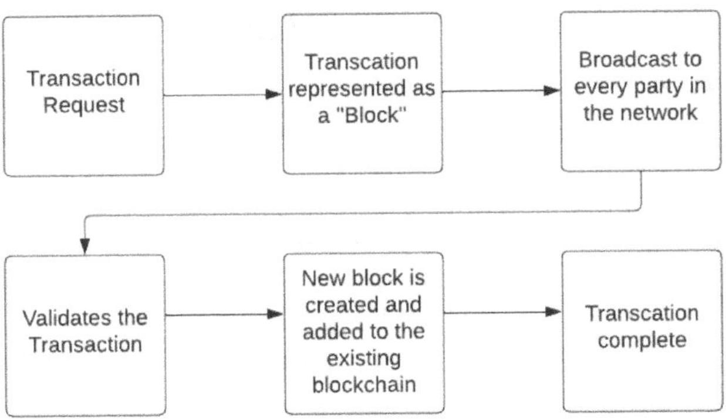

Figure 1. How blockchain works.

the name implies, organizes data into blocks that are strung together. This data structure creates an irreversible data chronology when implemented in a decentralized fashion.

Once filled, a block is permanent and added to the timeline. A precise timestamp is given to each block as it is added to the chain. Different types of blockchains are needed for various use cases. Public blockchains, private blockchains, consortium blockchains, and hybrid blockchains are the four basic types of blockchain networks. Each of these platforms has its own advantages, disadvantages, and ideal applications. This technology has several advantages and disadvantages as presented in Figure 1 because of its structural behavior.

For better clarity, the depth illustration about how blockchain functions is depicted in Figure 1 which shows that there are two parties: the sender and the receiver. When a transaction is requested, or when someone (A) wants to send money to someone (B) or make a transaction, it is first represented as a block. After that, the block is broadcasted to all the parties in the network, and those parties must all agree that the transaction is valid before a new block can be created and added to the blockchain. Finally, the transaction is completed, i.e., the money is transferred from A to B. The following terms explain some aspects of a blockchain.

Decentralization. Because no external organization is required to validate transactions or operations, transaction validation times are reduced [Dataflair team, 2018].

The network's distribution. Because no one controls the network when it's distributed, various users can always have several copies of the same data [Dataflair team, 2018].

Furthermore, because the failure of one node does not result in a network-wide failure, this feature makes the network durable.

Similarly, because information must be checked by multiple nodes in a distributed network, there are virtually no errors, making it nearly impossible to create erroneous or malicious information using blockchain technology.

Users pay a low price. Its decentralized structure enables speedy and easy transactions [Dataflair team, 2018].

These are some of the drawbacks of blockchain technology.

High implementation costs: Companies face high costs, which stymies mass adoption and implementation [Dataflair team, 2018].

Inefficiency. Multiple network users validating the same actions are inefficient because only one obtains the reward from the mining process [Dataflair team, 2018].

Keys that are only available to you. It is nearly impossible to retrieve private keys once lost, as has been recorded numerous times.

Storage. As the number of users grows, so does the number of operations that must be integrated into the blocks to be stored, necessitating a rise in the required storage capacity in the miners' computers [Dataflair team, 2018].

Unemployment. All intermediate processes for confirming payments and processes will inevitably be reduced due to eliminating intermediaries in the adoption and implementation of blockchain technology, resulting in an increase in the number of unemployed [Dataflair team, 2018].

3. Research Dimensions

1) Scripts called smart contracts are built on top of blockchain technology. They stand for a type of automation that allows the number of intermediary layers to be decreased or perhaps eliminated entirely. As a result, transaction, and enforcement costs, as well as processing times, are reduced by blockchain intelligent contracting systems. Additionally, we contend that smart contracts and blockchain can make cross-organizational collaboration easier. business operations. They can facilitate the integration of business owners and small and medium-sized enterprises into global supply chains by lowering high entry barriers and undermining the power of major players [Philipp Robert et al., 2019].

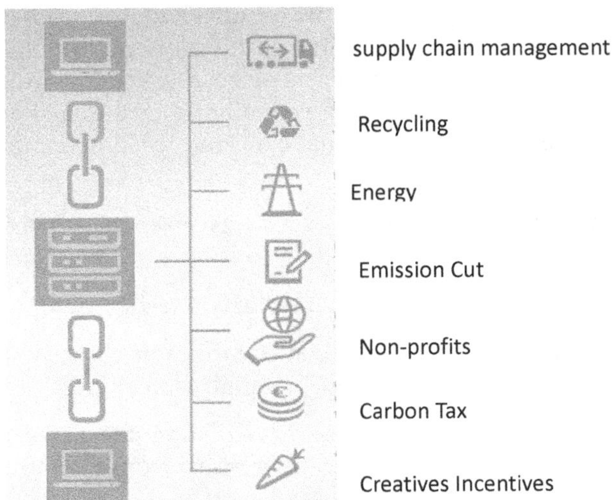

Figure 2. Blockchain role in environmental sustainability [Blockchain for climate action – DW – 05/08/2018].

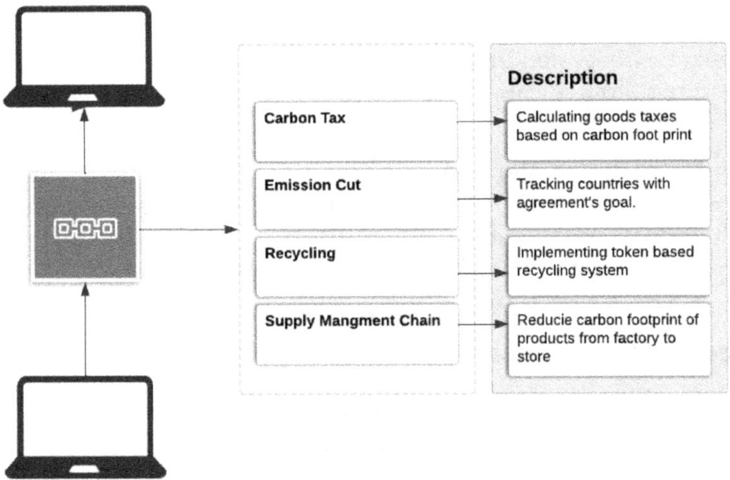

Figure 3. How blockchain works in environmental sustainability.

2) For effective energy consumption and environmental sustainability in blockchain-based IoT systems, they suggest a context-aware smart contract system using rule-based reasoning. The system gathers contextual data via smart contracts, analyzes it by considering relevant control rules, and then executes the necessary actions. All transaction history and logs are kept on the blockchain-based distributed ledger for immutable data provenance evidence, accountability, and traceability, ensuring data authenticity and integrity [Ngwira Lovemore et al., 2021].

Table 1. Types of Blockchain [Dataflair team, 2018].

	Private Blockchains	**Public Blockchains**	**Hybrid Blockchains**	**Consortium Blockchains**
Controlled by	✓ One authority, ✓ Permission required **[Dataflair team, 2018]**	✓ No central authority. ✓ No Permission required **[Dataflair team, 2018]**	✓ Single authority using a few processes without permission **[Dataflair team, 2018]**	✓ Controlled by group, ✓ Permission required. **[Dataflair team, 2018]**
Pros	✓ Access control, ✓ Performance **[Dataflair team, 2018]**	✓ Transparency, ✓ Trust **[Dataflair team, 2018]**	✓ Access control, ✓ Scalability **[Dataflair team, 2018]**	✓ Access control, ✓ Scalability **[Dataflair team, 2018]**
Cons	✓ Transparency, ✓ Trust **[Dataflair team, 2018]**	✓ Performance **[Dataflair team, 2018]**	✓ Transparency ✓ upgrading **[Dataflair team, 2018]**	✓ Transparency ✓ **[Dataflair team, 2018]**
Usage	✓ Supply chain ✓ **[Dataflair team, 2018]**	✓ Document validation **[Dataflair team, 2018]**	✓ Medical records **[Dataflair team, 2018]**	✓ Banking **[Dataflair team, 2018]**
Speed	✓ Fast **[Dataflair team, 2018]**	✓ Slow **[Dataflair team, 2018]**	✓ Slow/Fast **[Dataflair team, 2018]**	✓ Fast **[Dataflair team, 2018]**

Table 2. Types of Blockchain in healthcare and IoT.

References	**Pvt Blockchain**	**Public Blockchain**	**Hybrid Blockchain**	**Consortium Blockchains**	**Blockchain with Healthcare Techniques**	**Blockchain with IoT**
[Mondal et al., 2022]	x				x	
[Chen et al., 2021]	x					X
[Jain Apoorv et al., 2022]		X			x	
[Emad H. Abualsauod, 2022]			x			X
[Lin et al., 2022]				x		X

Table 3. Category of application based on their application.

Reference	**Category of Applications**	**Application —1 (Health care)**	**Application —2 (Education)**	**Application —3 (Energy)**	**Application —4 (Supply chain and Logistic)**
[Sun et al., 2022]	Health care	X			
[Badlani et al., 2022]	Education		x		
[Górski, 2022]	Energy			x	
[Radeva et al., 2022]	Supply chain and Logistic				X

3) Blockchain offers a supply chain control solution based on smart contracts to lower risks. Within the context of supply chain models, the technological capabilities of the proposed solution will be contrasted with those of a comparable technological approach, and their advantages and disadvantages will be assessed. As a result, the suggested method is appropriate for the dynamic management of intricate supply chains [Dietrich Fabian et al., 2020].

4. Dimensions of Environmental Sustainability

What is sustainability? Sustainability means attaining our aims without sacrificing the ability of future generations to accomplish their aspirations. We also require social and economic resources in addition to natural resources. Sustainability extends beyond ecological concerns.

Sustainable methods include solar, wind, hydroelectricity, and biomass energy. Crop rotation, crop cover, and wise water use are all aspects of sustainable agriculture, whereas selective logging and forest management are key components of sustainable forestry. Programs, projects, and actions targeted toward preserving a specific resource are generally considered sustainable. The four pillars of sustainability are human, social, economic, and environmental [**The four pillars of sustainability (futurelearn.com)**].

What is environmental sustainability? To support health and welfare both today and in the future, it is necessary to safeguard global ecosystems and conserve natural resources. Environmental sustainability seeks to enhance human well-being without unduly taxing the planet's life-sustaining ecosystems. Environmental sustainability aims to reduce pollution and harm to the environment while protecting natural resources and creating alternative energy sources. Given the number of resources we use daily, including food, energy, and manufactured goods, environmental sustainability is crucial. Due to the rapid increase in population, there has been an increase in greenhouse gas emissions from agriculture and manufacturing as well as unsustainable energy use and deforestation. Because of its sensitivity and importance, ecological sustainability is a word used in a variety of scientific domains. In today's world, there is far too much emphasis on environmental preservation in all sectors. Simply expressed, ecological sustainability refers to the preservation of natural resources and the protection of global ecosystems in order to support current and future health and well-being. Environmental sustainability is essential since we use so much energy, food, and artificial resources every day. Rapid population growth has resulted in increased greenhouse gas emissions, unsustainable energy use, and deforestation, all of which contribute to increased greenhouse gas emissions, unsustainable energy use, and deforestation. As a result, sustainability has to be supported by three pillars: economic, social, and environmental. Ecologically sustainable societies safeguard natural wealth and rely on its returns to survive, In any case, climate change is tied to environmental sustainability. As a result, it is critical to handle climate change in a multi-sectoral approach, as humans both cause and are affected by it. We need to assess a product's carbon footprint from cradle to grave to build a sustainable environment, which is known as product carbon foot printing (PCF). PCFs have proven themselves as a reliable method of estimating greenhouse gas emissions from goods and services across the supply chain. If we believe that measuring carbon footprints is done automatically, it is a costly and technically difficult task. However, if we follow the rules, we will have enough information about our surroundings to make informed decisions in the future. When it comes to environmental sustainability, the first step is to calculate the carbon footprint generated by various industries. In agriculture, PCF entails tracking greenhouse gas emissions from a variety of operations across the supply chain, including tillage, field plowing, pesticide application, harvesting, storage, and processing, as well as packing, transportation, and consumption. Data on the usage of energy and other resources, particularly water, is necessary for industrial production. Collecting on-site data (e.g., electrical appliances, outlets, and lamps) and using that data to respond correctly and produce a better consumption plan is crucial in the energy sector. People are curious about the influence of their decisions on climate change and desire knowledge about it. As a result, using blockchain technology to analyze Earth observation data would allow for more effective, efficient, and timely monitoring of environmental impacts and trends, as well as new insights into driving forces and environmental impacts and improved predictive capabilities. A process can also be verified against a set of emissions standards or the value of a specific environmental impact reduction measure. In general, understanding the principles of blockchain technology is necessary for judging its tremendous and transformative potential in society, the economy, and the environment.

5. Roles of Blockchain in Environmental Sustainability

Blockchain is undeniably a disruptive technology that is slowly making its way out of the banking and insurance niche and into the manufacturing, agri-food, and government administration sectors. The use of renewable energy can be expanded thanks to these systems. Sustainable and eco-friendly supply chain practices can also be implemented using blockchain technology. By making supply chains transparent, technology can follow products from the point of manufacture and assist in reducing inefficiencies and waste. As a result, the environmental impact of blockchain will be a hot topic in the coming years.

New prospects for green manufacturing, monitoring, and storing data-related activities that cause pollution and degradation and gathering and analyzing green or low-carbon data in real-time for prompt decision-making are all possible with blockchain technology. A measurable reduction in CO2 measured by an IoT-based network of atmosphere monitoring sensors placed around a village can "trigger" a blockchain smart contract, it provides a foolproof and cost-free mechanism for connecting favorable (or unfavorable) environmental changes or outcomes to monetary incentives or disincentives. We must examine things from various perspectives to have a better understanding.

Traditional blockchain systems necessitate the use of a significant amount of energy, which has a severe influence on the natural environment; for example, designing or manufacturing huge servers necessitates the construction of enormous buildings, which can negatively impact the landscape. It is best to look at the function of Blockchain in the environment from two different perspectives in this modern age and fast-increasing sector, as it has both positive and bad effects. We all know that if we use Blockchain in a restricted and efficient way, it can have a huge impact on numerous sectors of research.

A better policy can be proposed in the future if we have a greater understanding of the environment. We use a variety of blockchain technologies for these metrics, including measurement, analysis, and verification or validation. The data generated by these blockchain systems or technologies are devoid of any intermediary entities or parties, making it more reliable and, as a result, having several good consequences. When we perform studies using data from blockchain technologies, we have a good

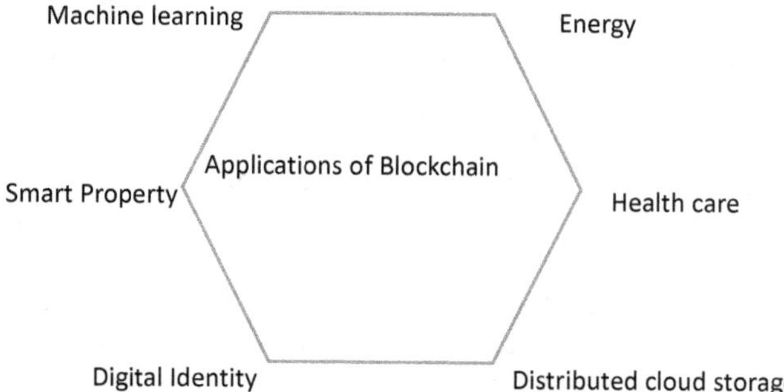

Figure 4. Potential application of blockchain technology [Blockchain Applications – Adaptive Learning and Optimization Lab (ufl.edu)].

chance of getting useful results. As a result, scientists will provide various ideas and outcomes for the community's well-being, and politicians may be able to propose various measures at the local and international levels, which might be highly essential for the world's sustainability, and their impact can be seen in a variety of industries, including agriculture.

In general, academics have emphasized the benefits of blockchain, but less attention has been paid to the environmental risks associated with its implementation. If we use blockchain technology, for example, we will need a vast number of servers to exchange data, goods, services, and information without the need for centralized authorities to authenticate identification and confirm transactions. These quick transactions or information transfers can be broadcast and synchronized digitally. Because of the carbon dioxide emissions emitted, these complex implementation problems contribute to environmental contamination, especially when handled on a big scale. Furthermore, the blockchain requires a significant amount of energy for crucial algorithms, processing, and calculations.

So, when discussing or implementing blockchain technology, we must consider both aspects, and if we want to do so on a large scale with sophisticated calculations, we must suggest a way to cut emissions and thus contribute to our environmental friendliness.

6. Basics of Contracts and Smart Contracts

A smart contract is computer software that executes corporate rules, commitments, and agreements that are often regarded as legal documents. In essence, smart contracts are scripts or software codes created by developers and placed on a blockchain, a phrase coined by cryptography pioneer Nick Szabo in 1994. They are expressed as transaction instructions, which are frequently triggered by events. Release payment to the supplier, for instance, if products are delivered to this customer's warehouse by this date. Therefore, Smart Contracts can carry out tasks automatically when businesses update their shipments and receipts. as a result Managing time-consuming and expensive manual business processes is no longer necessary.

An electronic program known as a "smart contract" automates the performance of business logic, commitments, and agreements.

An electronic warehouse receipt, bond, power units, currency units, futures contract, the share of risk, and many other things can all be represented by a smart contract [Cheng, S,2023].. Users on the network can produce, exchange, and settle these cryptographically distinct assets in real-time. Almost any form of business logic can be incorporated into a smart contract. According to the terms and circumstances of the contract, this business logic can be automatically enforced.

The contract responds to inputs by carrying out any duties or conditions that the contract's logic requires.

- A GPS coordinate confirming a ship's arrival at the proper port might cause payment to the seller of the commodities carried by that ship to be made automatically.
- A smart contract's ability to sell an option on a particular commodity may be activated by the input of that commodity's current price.
- If and when other requirements are satisfied, a buyer's signature on an invoice can create a payment obligation that is immediately executed on the date stated.
- Upon a default event as reported in the court filing system, collateral is transferred to the creditor.

As previously stated, Smart Contracts are often not binding contracts. They may, however, automatically execute agreements between parties. Additionally, as legal agreements frequently have a logical structure like code, such as if-this-then-that, paper-based agreements could be replaced by computer-based programs that automatically carry out a contract's conditions. As a result, Smart Contracts are crucial to the operation of blockchain models. It is possible to automate interactions between parties utilizing embedded smart contracts and automated rules, enabling parties to carry out their contractual obligations quickly, clearly, and effectively.

In essence, smart contracts are computer programs that automate the application of rules. Contrarily, traditional contracts are agreements that are written in plain, common language and are supported by legal obligations.

Smart contracts have the following qualities: they are tamper-proof because nothing can be changed after it has been coded; they are self-verifying due to automated possibilities; they are self-enforcing when the rules are always followed.

Simply put, smart contracts are blockchain-based algorithms that execute agreements when certain criteria are met. They are often used to automate the implementation of an agreement so that all parties can be certain of the conclusion right away, without the need for an intermediary or additional delay.

Smart contracts eliminate the need for a middleman, the legal system, or an external enforcement mechanism so that anonymous parties can carry out transactions and agreements. This eliminates any fees owing to these third parties and encourages frictionless transactions.

The code can either serve as the entire expression of the parties' agreement, or it can be used to supplement a typical text-based contract and carry out certain requirements, such as the transfer of monies from party A to party B. Because the code is duplicated over several nodes of a blockchain, it benefits from the blockchain's security, permanence, and immutability. This replication also means that the code is executed with each new block added to the network. The code will not execute any steps if no such transaction has been initiated. Most smart contracts are written in one programming language. directly suited for such computer programs such as solidity, Vyper etcA smart contract's objectives, its input parameters and execution phases must be specified clearly. To put it another way, if "x" occurs, perform step "y." As a result, smart contracts' actual tasks are simple, such as shifting a particular amount of cryptocurrency upon the fulfillment of specified conditions, from one party's wallet to another. Smart contracts are growing more complex and capable of conducting sophisticated transactions as blockchain becomes more ubiquitous and assets become more tokenized. Already, developers are integrating several transaction procedures to produce smart contracts that are increasingly complex. However, it will be many years before the code can assess more subjective legal criteria, such as whether a party has made commercially reasonable efforts or whether an indemnification provision should be triggered, and compensation paid.

Another step is required before a smart contract may be performed on certain blockchains, namely the payment of a transaction fee to add the contract to the chain and execute it. Smart contracts are now best suited for automating two types of "transactions' seen in many contracts: (1) ensuring payment of cash in response to specific triggering events, and (2) imposing financial fines if requirements are not satisfied. In any case, once the smart contract is set up and running, humans will be able to interact with it.

The financial industry can use smart contracts for lending, borrowing, and investing. They can be used in real estate, healthcare, gaming, and even the configuration of entire company structures.

7. Applications and Techno Convergence of Blockchain and Smart Contract

The development and operation of cryptocurrencies is the most well-known application of blockchain (e.g., Bitcoin, Ether and others). Blockchain is being considered in other sectors besides financial services, such as international trade, taxation, supply chain management, corporate operations, and governance. Researchers have used blockchain technology in numerous fields, such as supply chain management, security, and privacy, edge computing and artificial intelligence. Although blockchain is still in its early stages, experts agree that it has the potential to help organizations meet the demands of the fourth industrial revolution. Blockchain is expected to significantly alter numerous fields of studies. The appeal of blockchain arises from its capacity to facilitate transparent data interchange, streamline company processes, minimize operational costs, improve collaboration efficiency, and establish a system that does not require explicit faith in its control, as supply chains do.

Among its many advantages, blockchain can help to solve a variety of environmental sustainability issues by promoting environmental sustainability through three essential foundational mechanisms: resource rights, product provenance, and behavioral incentives. It could open new possibilities for green manufacturing and the creation of green supply chains., as well as real-time collecting and analysis of green or low-carbon data for rapid decision making and the creation of green supply chains.

Many academics who have studied blockchain technologies have concluded that this technology can help create sustainable, green supply chain monitoring and reduce CO2 emissions by monitoring the exchange of hazardous waste and creating an incentive system that encourages recycling, improves circular economy practices, and monitors the use of natural resources.

The role of blockchain in creating a sustainable society was recently highlighted, with an explanation of how different digital blockchain solutions can support sustainability from three perspectives, depending on which topic the technology can be targeted: service delivery, resource management, and city governance.

Blockchain applications are being developed using a peer-to-peer logic in which organizations can exchange goods, services, and information without the need for central bodies to verify identity, validate transactions, or enforce commitments, or at the very least without the need for as many intermediaries as currently exists. At a basic level, this might lead to increased efficiency and cost savings for businesses and other groups.

There are several benefits to adopting blockchain technology for many applications. For example, using block technology for smart contracts can be quite beneficial. There are advantages and disadvantages to using a block technology system to handle contract-related difficulties. If we look at the problems from multiple perspectives, let's take reliability. As a feature, blockchain technology allows for the exchange of goods, services, and information between organizations without the use of a middleman. As a result, during this exchange, senders, and recipients as well as buyers and sellers can exchange what they need without the middleman's interference, resulting in a high probability of accuracy interchange of information. As the number of users increases, so do the processes that must be integrated into the blocks to be stored, increasing the amount of storage space needed in the computers used by the miners.

8. Summary and Conclusions

The structure of this chapter is as follows: The first part attempts to describe Blockchain, its evolution, types, pros, and cons. We suggest various study papers on blockchain and its use in part two. In parts three and four, we primarily discuss how blockchain technology is used for environmental sustainability and its role in sustainability issues. Part five explains the fundamentals of contracts and smart contracts, and the part after that discusses applications and the technological convergence of blockchain and smart contracts. We conclude with a summary here.

Blockchain technology offers operational and regulatory benefits, as well as increased transparency, traceability, and verification capabilities. This technology is also a strong database that can be easily coupled with Big Data. Blockchain technology has the potential to reduce the cost and increase the competitiveness of many services, such as finance, healthcare, law enforcement, Internet of Things, and environmental sustainability. In addition, there are various other industries where blockchain technology is used.

References

Ahad, M.A., Paiva, S., Tripathi, G. and Feroz, N. (2020). Enabling technologies and sustainable smart cities. Sustainable Cities and Society, 61: 102301. [online] Available from: Enabling technologies and sustainable smart cities - ScienceDirect.

Alia Al Sadawi, Batool Madani, Sara Saboor, Malick Ndiaye and Ghassan Abu-Lebdeh. (2021). A comprehensive hierarchical blockchain system for carbon emission trading utilizing blockchain of things and smart contract. Technological Forecasting and Social Change, 173: 121124.

Ante Lennart. (2021). Smart contracts on the blockchain—A bibliometric analysis and review. Telematics and Informatics, 57: 101519.

Available from The four pillars of sustainability (futurelearn.com), The four pillars of sustainability.

Badlani, S., Aditya, T., Maniar, S., Devadkar, K., EduCrypto: Transforming Education using Blockchain. (2022). Proceedings—2022 6th International Conference on Intelligent Computing and Control Systems, ICICCS 2022, pp. 829–836.

Blockchain Technology. Advantages & Disadvantages of Blockchain Technology. [online]. (2016). Available from Advantages & Disadvantages of Blockchain Technology – Blockchain Technology (wordpress.com).

Bualsauod, Emad H. (2022). A hybrid blockchain method in internet of things for privacy and security in unmanned aerial vehicles network. Computers and Electrical Engineering, 99: 107847. https://doi.org/10.1016/j.compeleceng.2022.107847.

Chen, X., Nguyen, K. and Sekiya, H. (2021). An experimental study on performance of private blockchain in IoT applications. Peer-to-Peer Netw. Appl., 14: 3075–3091. https://doi.org/10.1007/s12083-021-01148-9.

Cheng, S. (2023). Metaverse: Concept, Content and Context from https://books.google.com.et/books?id=DubGEAAAQBAJ, 2023, Springer Nature Switzerland.

Christine Parizo. (2021). What are the 4 different types of blockchain technology?. [online] .Available: What are the 4 different types of blockchain technology? (techtarget.com).

Francesca Dal Mas, Grazia Dicuonzo, Maurizio Massaro and Vittorio Dell'Atti. (2020). Smart contracts to enable sustainable business models. A case study. Management Decision.

Dataflair Team. (2018). Advantages and disadvantages of Blockchain Technology [online]. 2018. Available from Advantages and Disadvantages Of Blockchain Technology - DataFlair (data-flair. training).

Fabian Dietrich, Ali Turgut, Daniel Palm and Louis Louw. (2020). Smart contract-based blockchain solution to reduce supply chain risks. IFIP International Conference on Advances in Production Management Systems. Springer, Cham.

Górski, T. (2022). Reconfigurable smart contracts for renewable energy exchange with re-use of verification rules. Applied Sciences (Switzerland), 12(11): 5339.

Hafiza Yumna, Muhammad Murad khan, Maria Ikram, Sidra Noreen and Romesa Sabeen. (2019). Use of blockchain in Education: A Systematic Literature Review [online]. Available: Use of Blockchain in Education: A Systematic Literature Review | SpringerLink.

Hewa Tharaka, Mika Ylianttila and Madhusanka Liyanage. (2021). Survey on blockchain-based smart contracts: Applications, opportunities, and challenges. Journal of Network and Computer Applications, 177: 102857.

Jain Apoorv Tripathi and Arun Kumar. (2022). Blockchain enable smart contract-based telehealth and medicines system. Communications in Computer and Information Science, 1591 CCIS, pp. 724–739.

Li Jennifer and Mohamad Kassem. (2021). Applications of distributed ledger technology (DLT) and Blockchain-enabled smart contracts in construction. Automation in Construction, 132: 103955.

Lin, T., Huan, Z., Shi, Y. and Yang, X. (2022). Implementation of a Smart contract on a consortium Blockchain for IoT Applications. Sustainability (Switzerland), 14(7): 3921.

Weizhi Meng, Jianfeng Wang, Xianmin Wang, Joseph Liu, Zuoxia Yu, Jin Li, Yongjun Zhao and Sherman S.M. Chow. (2018). Position paper on blockchain technology: Smart contract and applications. International Conference on Network and System Security. Springer, Cham, 2018.

Mondal, S., Shafi, M., Gupta, S. and Gupta, S.K. (2022). Blockchain based secure architecture for electronic healthcare record management. GMSARN International Journal, 16(4): 413–426.

Szabo, N. (1997). The Idea of Smart Contracts [online]. Available: Nick Szabo—The Idea of Smart Contracts (uva.nl).

Lovemore Ngwira, Mpyana Mwamba Merlec,Youn Kyu Lee and Hoh Peter In. (2021). Towards context-aware smart contracts for blockchain IoT Systems. International Conference on Information and Communication Technology Convergence (ICTC). IEEE, 2021.

Oberhauser Daniel. (2019). Blockchain for environmental governance: Can smart contracts reinforce payments for ecosystem services in Namibia? Frontiers in Blockchain, 2: 21.

Philipp Robert, Gunnar Prause and Laima Gerlitz. (2019). Blockchain and smart contracts for entrepreneurial collaboration in maritime supply chains. Transport and Telecommunication, 20.4: 365–378.

Radeva, I. and Popchev, I. (2022). Blockchain-enabled supply-chain in crop production framework. Cybernetics and Information Technologies, 22(1): 151–170.

Rejeb Abderahman and Karim Rejeb. (2020). Blockchain and supply chain sustainability. Logforum, 16.3.

Roberto Leonardo Rana, Pasquale Giungato, Angela Tarabella and Caterina Tricase. (2019). Blockchain applications and sustainability issues [online]. Available from CEEOL - Article Detail.

Sara Saberi, Mahtab Kouhizadeh, Joseph Sarkis and Lejia Shen. (2019). Blockchain technology and its relationships to sustainable supply chain management. International Journal of Production Research, 57.7: 2117–2135.

Saef Ullah Miah, M., Mashiour Rahman, Md. Saddam Hossain and Aneem Al Ahsan Rupai. (2019). Introduction to Blockchain.

Stark, J. (2016). Making Sense of Blockchain Smart Contracts [online]. Available: Making Sense of Blockchain Smart Contracts - CoinDesk.

Sun, Z., Han, D., Li, D., Wang, X., Chang, C.-C. and Wu, Z. (2022). A blockchain-based secure storage scheme for medical information. Eurasip Journal on Wireless Communications and Networking, 2022(1): 40.

Types of Blockchain [online] Available from What Are The Different Types of Blockchain Technology? - 101 Blockchains.

Wright Craig and Antoaneta Serguieva. (2017). Sustainable blockchain-enabled services: Smart contracts. 2017 IEEE International Conference on Big Data (Big Data). IEEE, 2017.

What is Blockchain Technology? What is Blockchain Technology? - IBM Blockchain | IBM.

Use cases of blockchain technology in business and life (2022). [online]. Available Use Cases of Blockchain Technology in Business and Life (insiderintelligence.com).

Chapter 4

Blockchain and its Practical Implications in Waste Management Companies

A Practitioners' View

Secinaro Silvana,[1] *Calandra Davide*[1,*] and *Marseglia Giuseppe Roberto*[2]

1. Introduction

Awareness of the importance of environmental sustainability and waste management has increased in recent decades. Population growth and economic development have led to increased waste generated and the need to find sustainable solutions to manage it (Richter et al., 2021). Waste management companies play a crucial role in this context. Their activities include collecting, transporting, treating and disposing waste to minimize environmental impact and promote sustainability and circularity (Jabbour, 2013). To achieve this goal, waste management companies must adopt sustainable practices and seek to minimize the use of natural resources, the emission of pollutants, and the generation of non-recyclable waste (Calandra et al., 2022). In addition, these companies can

[1] Department of Management, University of Turin, Italy.
[2] Department of civil engineering and architecture. University of Pavia, Italy.
Emails: silvana.secinaro@unito.it; r.marseglia@aliaserviziambientali.it
* Corresponding author: davide.calandra@unito.it

also promote a culture of sustainability by educating and raising public awareness of waste management issues and reducing environmental impacts. Adopting sustainable practices by waste management companies promotes environmental sustainability and can lead to economic benefits. Reducing the costs associated with waste management can increase the profitability of companies and improve their public image (Abubakar et al., 2022).

Today, technology has an increasingly significant impact on waste management, offering new solutions to address challenges related to environmental sustainability. Emerging technologies such as artificial intelligence, IoT, blockchain, robotics, and sensors are revolutionizing waste management, making it more efficient, sustainable, and transparent. Using these technologies, waste management companies can monitor the waste supply chain, improve the planning of waste collection and management activities, predict the amount of waste produced in the future, automate some waste management activities, increase the transparency and traceability of the waste management process, and incentivize citizens to participate in waste management actively (Esmaeilian et al., 2018).

The application of blockchain is revolutionizing the waste management industry by offering innovative solutions to address challenges related to environmental sustainability (Salah et al., 2019). Due to its decentralized and secure nature, blockchain allows for tracking the waste supply chain, monitoring the recycling process, managing carbon credits and environmental offsets, creating new sustainable business models, and incentivizing collaboration among public sector companies (Sahoo et al., 2022). Despite the growing interest in using blockchain in waste management, studies still need to explore its effectiveness and ability to promote environmental sustainability. Therefore, the study aims to answer the following question:

RQ: How can blockchain improve waste management companies' activities?

It attempts to explore in depth the use cases of blockchain in waste management and assess their impacts on environmental sustainability; analyze the challenges and opportunities related to implementing blockchains in waste management, such as data privacy, security, cost, and scalability; examine the social, economic, and environmental implications of implementing blockchain in waste management; and assess its impact on local communities and regional economies. Identifying the research problem and gap may help researchers develop a clear and detailed research plan to deepen knowledge of the use of blockchain in waste management. It may also help develop practical solutions to address environmental sustainability challenges, bridging the possible mismatch between what is already in use and what is expected by practitioners (Romme et al., 2015).

To address our research aim, we conducted empirical research using thematic analysis. Our results show experiences, case studies and theories related to use of blockchain in waste management companies. Specifically, we uncover some experiences related to tracking the waste supply chain, improving transparency and traceability of the waste management process, incentivizing collaboration between public and private companies in waste management, and creating new sustainable business models.

The implications suggest the need to promote the adoption of blockchain technology in waste management and to address the challenges related to its implementation. In addition, the study suggests that blockchain can promote environmental sustainability and incentivize public and private companies to collaborate in waste management.

2. Methodology

2.1 Thematic and Content Analysis

This section adopts a qualitative methodology. According to Krippendorff (2013), written and multimedia material may be investigated using qualitative approaches such as thematic and content analysis. Content analysis allows a thorough study of different research streams (Hsieh and Shannon, 2005). It is used in several fields to highlight certain literature trends and topics and look for new and relevant information (Landrum and Ohsowski, 2018). Additionally, according to Massaro et al., 2021, content analysis may be used to investigate practitioner results and perspectives. However, content analysis has also been used for other purposes. It may be used to analyze how firms disclose information as a case study or to comprehend intellectual capital knowledge within specific literature (Ruzza et al., 2020). Or, as suggested by Secinaro et al., 2021, it is possible to investigate and gather practitioners' experiences and case studies using content analysis. Therefore, considering the aforementioned elements, content analysis appears to be a methodology allowing for conventional, direct, and summative scientific research (Hsieh and Shannon, 2005).

Additionally, this approach may be combined with an in-depth thematic analysis. As suggested by Clarke et al. (2015), a thematic analysis may add relevant details for complex processes or research fields, giving a holistic understanding of narrative elements and cases. Consequently, following the literature review on waste management and the new and disruptive opportunities from new technologies, this study aims to share knowledge and give practical tools and case studies to managers adopting the blockchain in this field.

2.2 *Data Collection and Analysis*

In addressing our research question and aims, we use the NexisUni database to collect practitioners' sources. According to Massaro et al. (2021) and Smith et al. (2021), this database features more than 17,000 news, business and legal sources. Therefore, it allows information acquisition from blogs, newspaper reports, and magazines to gather practical experiences and business opportunities.

The analysis was conducted in April 2023 using the following search string, connected by Boolean operators:

TITLE-ABS-KEY "Blockchain" AND "Waste Management".

The first stage revealed 2,020 results. Subsequently, the authors restricted the search considering the last two years and validating only newswires, press releases, web-based publications and business opportunities. Additionally, we excluded news related to the stock market and market value. The choice follows previously published research by Massaro et al., 2018; Secinaro et al., 2021. This research strategy allowed us to extract 324 sources.

In the final stage, we used the Leximancer software. It uses automated content analysis to search for and extract thesaurus-based ideas from text data (Smith and Humphreys, 2006). According to the previous authors, the software allows for extracting co-occurrence, semantic, and relational information. As a result, using Leximancer enables the mapping of links between blockchain and waste management ideas and case studies. The study utilised its technological characteristics to open and read each extracted document by coding the content and using an Invivo technique (Miles et al., 2014).

3. Results

3.1 *Blockchain in Waste Management Companies: Case Studies and Applications*

The following section explains practitioners' themes investigating the link between blockchain and waste management, focusing on theories, applications, and case studies. As shown in Figure 1, the main subject of blockchain is linked to many other concepts such as "waste", "technology", "digital", "sustainable", "recycling", "carbon", "future", "market", "supply", "plastic" and "crypto".

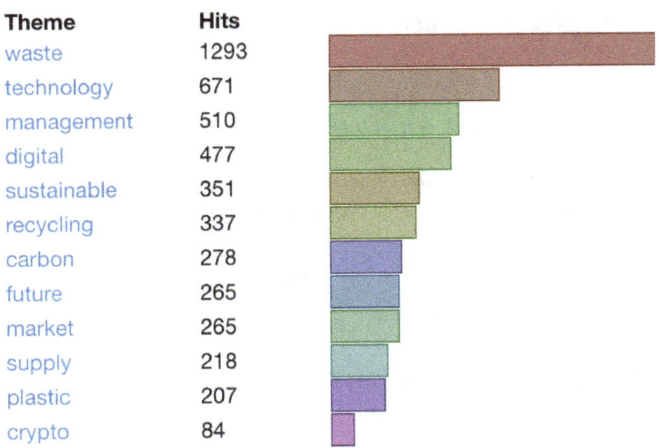

Figure 1. Themes under practitioners' discussion (Source: Authors' elaboration).

The first research topic considers the waste management sector's challenges in addressing sustainability needs. Other research topics consider the analysis of new digital solutions to map and account for new digital solutions that enhance citizens' engagement in waste management processes. Finally, a futuristic element from the analysis of practitioners' sources, considering the role of cryptocurrencies in enhancing recycling processes, is also linked with the role of companies in this sector. More elements can be found in Figure 1.

Among the most relevant themes, several recurring keywords emerge that identify cases and experiences of blockchain use in this field of study. As shown in Figure 2 and Table 1, the first area of research we discover concerns materials recycling. For example, Motshweni (2022) described Project Up as an initiative that, through blockchain technology, aims to connect citizens as waste contributors and recyclers on the one hand, and connect buyback centers on the other. The technology enables and incentivizes material recovery, enabling transactions to track and convert what is recycled into coupons or rewards (Calandra et al., 2022). Therefore, blockchain, through data verifiability and control, can automate recycling transactions without intermediaries and provide direct facilitation to citizens by activating elements such as awareness and concern toward environmental aspects (Bigliardi et al., 2020; Guo et al., 2020).

A second experience of using blockchain for recycling comes from Southeast Asia. For example, the company AIS of Thailand (AIS, 2022) has launched a new business model on the market that aims to create an ecosystem to dispose of e-waste. Bringing e-waste such as a used smartphone, tablet, or pc to a recycling center allows the citizen to check and calculate the greenhouse gas emissions generated. Specifically, the

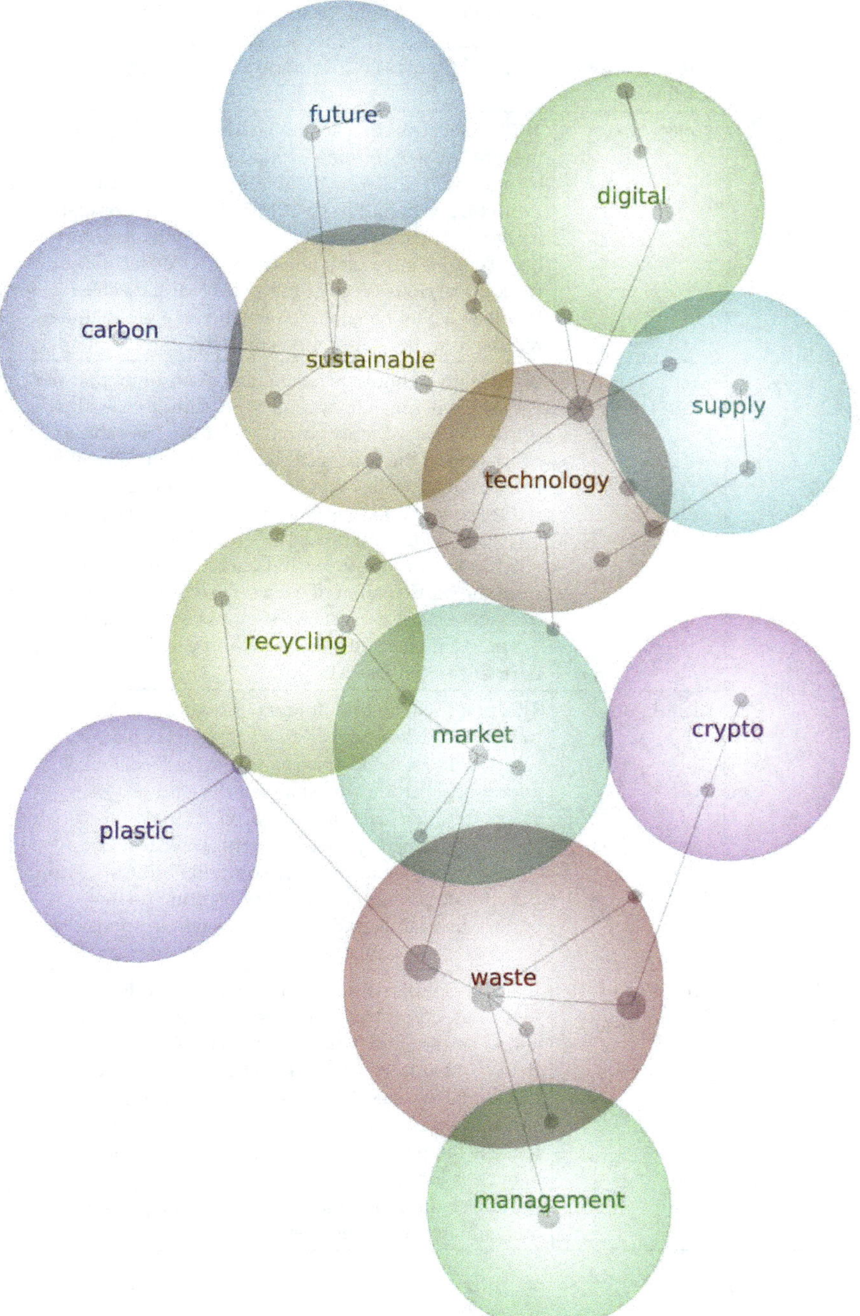

Figure 2. Thematic analysis (Source: Authors' elaboration).

Table 1. Practitioners' debate on Blockchain for waste management.

Macro Topics	Themes/Case studies	Original Quotes and Reference
Recycling	Project Up	This project uses blockchain technology to improve the efficiency of reclaimers and buy-back centres by making it easier for them to record transactions. Project Up is an initiative of PET Recycling Company, a provider of sustainable solutions. It will allow buy-back centres and reclaimers to record transactions and track the various types of materials that they collect. It will also help reclaimers track the revenue that they receive from the transactions (Motshweni, 2022).
	AIS	Blockchain technology has made us confident that the E-Waste will make its way to a recycling process that is up to standard, transparent, and fully traceable at every step of the process. It has also become possible to calculate Carbon Scores as an indicator of efforts to reduce greenhouse gas emissions which can be shared on social media and be identified in the Metaverse. AIS has a plan to develop the Carbon Scores arising from the participation of each sector for the correct management of E-Waste, so they can be used as Utility Tokens to further the businesses of participating partner companies (AIS, 2022).
	KleanLoop	This plant will be fully integrated with Klean's proprietary KleanLoop™ Blockchain SaaS platform which will track and trace all incoming feedstock, operations, carbon emissions offsets, and product sales offering a level of transparency not seen before in the Philippines recycling industry. The KleanLoop™ platform is a key technology solution built for solving the climate change crises and preventing further environmental degradation in a socially responsible society (Klean Industries, 2022).
	Ecotrace Solutions	Ecotrace Solutions, a Brazilian startup offering blockhain-based traceability tech for the agrifood value chain, has secured 3 million reals ($577,000) in funding from São Paulo-based VC firm KPTL. Renato Ramalho, CEO of KPTL, said the objective of the new fund "is to serve, in this order, 'outside the fence' and 'inside the fence'. Having strong attributes towards to the ESG agenda is mandatory. The main areas are the animal protein chain, grains, sugar-alcohol, biotechnology, and climate (AgFunderNews, 2021).

Table 1 contd. ...

...*Table 1 contd.*

Macro Topics	Themes/Case studies	Original Quotes and Reference
Carbon Accounting	Hydrogen transportation	For example, the use of hydrogen in transportation is an emerging market and can lower carbon emissions as well as decarbonize Scope 3 emissions across value chains. However, green hydrogen comes with the additional challenge of tracing/guaranteeing its origin. Ensuring green hydrogen was produced with electricity from renewable energy or with low carbon-emission fuel for which CO_2 emissions are captured and stored is essential. A blockchain-based platform can collect data from IoT-monitored energy plants to verify the share of renewable energy injected in the electrolysis and the share of CO_2 captured in the case of fossil fuel energy. The platform will record the amount of hydrogen produced in the process and monitor the delivery of the hydrogen. The audit trail on the blockchain ensures that all the information shared in the platform is transparent and tamper-proof (World Economic Forum, 2023).
	Phantasma	Phantasma is proud to announce that its blockchain is certifiably carbon negative. Building its industry-leading platform and tools with its sustainable, green future in mind, Phantasma's carbon footprint is miniscule by design. However, carbon footprints — no matter how small — are not something anyone wants to leave behind for future generations to deal with. Phantasma's partner Save Planet Earth is dedicated to lessening the impact of global warming. Since its inception, Save Planet Earth has built up carbon sequestration initiatives around the world, including approval of the planting of over 100 million trees, initiatives in solar, hydro, wind, beach/ocean clean-up, recycling, waste management and more. These initiatives generate certified carbon credits that can be provided to entities wishing to offset their carbon footprint. Phantasma enables Save Planet Earth to deploy its Carbon Credit NFTs (CCNFTs) on a certified carbon negative blockchain while Phantasma is able to acquire home-grown CCNFTs offsetting its carbon footprint and ensuring that the infrastructure and tooling it provides does not adversely affect the shared planet (GlobeNewswire, 2022).

Table 1 contd. ...

...Table 1 contd..

Macro Topics	**Themes/Case studies**	**Original Quotes and Reference**
Sustainable Business models	Satma CE	Hasiru Dala Innovations, the India-based social impact organisation that works with waste pickers and other waste workers to ensure a life with dignity while managing all municipal solid waste streams has implemented Satma CE (scan and trace material application across the circular economy),– a cloud SaaS software that also uses blockchain to track the movement of waste from the point it enters the recycling supply chain through waste-picker entrepreneurs or scrap dealers to the final manufacture of new packaging or products that use this very waste (Cision PR Newswire, 2022).
Crypto currencies	Solar Coin	Offsets allow emitters to mitigate emissions from a carbon-accounting standpoint. Crypto enthusiasts are driving up demand for carbon offsets in a new wave of ReFi. Toucan aggregates carbon offsets from various established registries like Verra (off-chain registries) and converts them into carbon tokens on one blockchain super-registry to increase transparency, verifiability and asset liquidity. KlimaDAO took the concept a step further, attempting to sweep all offsets from off-chain registries into one pool to create artificially scarce offsets that would drive up the price of offsets. Similarly, new cryptocurrencies like bitcoin zero and SolarCoin are emerging with a unique value proposition – owning them means owning a unit of the underlying currency and a unit of carbon offset (Singh et al., 2022).
Public management	United States Environmental Protection	Using EPA funding, UIC-Chicago will partner the Solid Waste Agency of Lake County, Illinois, to develop and demonstrate Blockchain Environmental Asset Tracking to manage solid waste and recyclable materials (US EPA, 2021).

Source: Authors' elaboration.

ecosystem allows anyone (an individual or an organization) to manage e-waste and, through a *"Track and Trace"* mechanism, know the status of each recycled piece throughout the process and calculate how much Co_2 emissions have been saved.

As defined by Saichon Submakudom (Head of Public Relations): *"In the global context of digital tech becoming part of everyone's lives, digital devices have proliferated. At AIS, we have taken up two key missions: first, to build awareness about the dangers of E-Waste, and second, to get involved in processes to collect and dispose of E-Waste correctly. We have been working with partners to set up drop-off points, from where waste is taken to the recycling process"*. Therefore, in this experience, we discover how technology can raise awareness about environmental issues and especially generate involvement in properly collecting and disposing waste (Kewell et al., 2017).

Also, in this area, we discover the case of KleanLoop. The company claims that (Klean Industries, 2022), this is the first case of tire recycling in North America. Specifically, the case reported here demonstrates cooperation between a recycling company and a digital application programming and management company. As indicated by the press releases, blockchain enables monitoring of the reintegration into supply chains of recycled tires by allowing tracking of all raw materials entering, leaving, and any offsets to carbon emissions saved (Hua et al., 2020). In addition, as indicated by the CEO of Klean Industries Inc, Jesse Klinkhamer: "*With an increasingly regulated industry, data is the only way we can solve the global waste crisis. The industry needs true and accurate data that we can rely on. It's critically important that the industry becomes more compliant and sustainable, and the KleanLoop™ is the system we need that provides transparency throughout the entire supply chain. We are confident that this innovation is the solution that can and will achieve that*". Therefore, through blockchain and data, it is possible to improve the aspects required to increase environmental sustainability (Bag et al., 2021).

Finally, through the case of Ecotrace Solutions, we discover a case related to the agribusiness sector (AgFunderNews, 2021). Through a combination of Blockchain, Machine Learning and the Internet of Things, the company succeeds in classifying cattle and assigning each head of cattle a barcode to monitor a range of information regarding the environmental conditions relating to growth, using a framework about environmental sustainability at the stages of growth. Finally, through a technology mix, one can raise awareness about the sustainability criteria by which the cattle have grown over time (Patelli and Mandrioli, 2020).

The second macro-theme concerns carbon accounting. Here the evidence discovered is different and belongs to multiple areas. For example, the World Economic Forum, 2023 suggests using blockchain and other technological systems to ensure that hydrogen in the mobility sector to reduce Scope 3 carbon emissions across value chains. The tracking challenge enables user awareness and prioritizes transparency in using alternative energy from fossil fuels. This is close to the aforementioned case (AIS, 2022). Also, in that context, blockchain is used to increase confidence in e-waste recycling and enable the calculation of Carbon Score as a rating for tracking Co_2 produced. Finally, blockchain enables action by users who, thanks to a saving of emitted Co_2, can use 'Utility Tokens' to obtain discounts on products and activities. An additional case study discovered involves Phantasma (GlobeNewswire, 2022). In this case, the mission is about the opportunity to measure the carbon footprint of individuals and companies. By evaluating lifestyle habits, users can ascertain their footprint in terms of Co_2 and decide to offset it through activities such as tree planting, solar, hydroelectric, and wind energy initiatives, beach and ocean cleanups, recycling, and waste management. These credits can be distributed among stakeholders through specific tokens (CCNFT).

The third macro-theme discovered concerns sustainable business models (SBM). For example, Hasiru Dala Innovations is an Indian waste management organization that has launched a new platform based on blockchain technology that enables tracking of materials. In addition, through the Software as a Service (SaaS) cloud, the company provides tools for entrepreneurs to manage corporate waste, paving the way for a new business model and enabling environmental sustainability vis-à-vis stakeholders (Cision PR Newswire, 2022).

The fourth macro-theme concerns cryptocurrencies. As indicated by Singh et al., 2022, blockchain enables tracking environmental offsets and creating green carbon accounting. In addition, the ledgers created allow for the exchange and creation of trading markets for green credits and Co_2 offsets. Therefore, units of carbon offsets often derived from industrial activities can be rewarded in crypto currency.

Finally, the fifth macro-theme concerns the role of public companies. The Chicago EPA, a waste management company, is an example where blockchain was launched directly to monitor and manage municipal solid waste by virtual monitoring the recycling of materials used (US EPA, 2021).

4. Discussion and Conclusion

This section aims to compare practitioners' sources regarding the open debate on the role of blockchain in waste management. To this end, the research team adopted a mixed qualitative approach based on thematic and content analysis, extracting exemplary experiences to guide managers' future choices.

The qualitative analysis identified five macro themes. We discover a strong interest in blockchain applications in waste management as seen from several sources (AIS, 2022; Gatteschi et al., 2018; Motshweni, 2022; Scekic et al., 2019).

At the same time, we discover a debate on the enhancement of the technology itself to improve environmental sustainability. In particular, as discussed at the academic level (Bakarich et al., 2020), blockchain will be able to increase the quality and reliability of data with many implications for practice. Therefore, blockchain will be able to provide factual accounting support to enable the inclusion of multiple stakeholders and ensure that there are uniform standards in the assessment of sustainability reports. From a practical perspective, these assumptions are confirmed by the cases of United States Environmental Protection (US EPA, 2021), AIS (AIS, 2022) and KleanLoop (Klean Industries, 2022). Such concepts recall the collaborative design theory. As indicated by Grushina, 2017, communicative dialogue requires a specific collaborative design between organization and stakeholders, especially in corporate social responsibility. So, technology-mediated engagement and collaboration enable certification of data in the cycle's context and increase the reporting quality.

In parallel, practitioners urge us to consider some challenges concerning green accounting and their connection with new cryptocurrencies. As sensitized by the World Economic Forum, 2023 and applied by Phantasma (GlobeNewswire, 2022), blockchain creates a digital ledger for Co_2 reporting while enabling offset trading. Despite primary enthusiasm, this phenomenon appears to be under investigation in academia. For example, scholars Mohsin et al. (2023) discover how cryptocurrencies can be an inverse phenomenon to environmental sustainability. In fact, in high-mining countries, the more virtual currencies are traded, the greater the consumption of e-waste energy and increased emissions.

Finally, the links between theory and practice in the context of Sustainable Business Models (SBM) are interesting. As defined by Bagnoli and Maura, 2021; Biloslavo et al., 2018, social and environmental needs make customers increasingly sensitive. Therefore, new business models become necessary, paving the way for new business opportunities for companies. As suggested by Calandra et al., 2022, blockchain, through its features of decentralization, verifiability, and decentralization enables new archetypes in different sectors and encompassing the different areas of environmental, social, and economic sustainability. Our results align with this vision and are confirmed by the case of Satma CE, which started a new technology-based business line from a waste management company to provide recycling monitoring solutions.

Considering the experiences of practitioners, we show how blockchain can be used to support waste management companies and raise awareness among those who deliver waste by mapping the entire recycling value chain. It is interesting to note how the practical aspects have direct links to the theoretical approaches outlining, in fact, a dedicated research strand.

Our study presents some theoretical implications. The thematic analysis by extracting practical cases allows us to initiate a discussion about blockchain observed through collaborative design theory. Therefore, the theory can support identifying key variables such as trust, transparency and data accessibility. In addition, our contribution confirms the role of technology in enabling sustainable business models in the three areas, environmental, social and economic.

Our study also has interesting practical implications for managers of waste companies. The thematic research of the results allowed us to discover multiple applications encapsulating the areas of recycling, green accounting for Co_2 monitoring, launching new sustainable business models, the emergence of dedicated cryptocurrencies, and the role of collaboration with other public companies. Therefore, managers can apply blockchain or search for solutions that can increase the value of business activities and make people aware of the importance of climate change and sustainability on the environmental side. However, as discovered, they must ensure the technology is not an environmental drawback. So, they will have to ensure the proper use of blockchain by limiting its negative impact.

4.1 Limitations and Future Research Perspectives

Our study, like any research, has some limitations that can be seen as research opportunities. The first is related to search and keyword usage. In particular, future studies could better understand the evolution of blockchain in the waste management sector through more structured searchkeys. The second limitation relates to the database and sources used. Again, further future research could also combine and observe the views of academics and stimulate a debate enriched by publication data, themes, methodologies and theories. Finally, the survey was conducted using a qualitative methodology to validate practitioners' sources. This paves the way for future research that can adopt quantitative and qualitative methods to investigate exploratory and interventionist case studies and verify blockchain's application and user perceptions in household and business waste management.

References

Abubakar, I.R., Maniruzzaman, K.M., Dano, U.L., AlShihri, F.S., AlShammari, M.S., Ahmed, S.M.S., Al-Gehlani, W.A.G. and Alrawaf, T.I. (2022). Environmental sustainability impacts of solid waste management practices in the Global South. Int. J. Environ. Res. Public Health, 19, 12717. https://doi.org/10.3390/ijerph191912717.

AgFunderNews. (2021). Brazilian blockchain startup Ecotrace gets funding for traceability tech. Retrieved April 16, 2023, from AgFunderNews website: https://agfundernews.com/brazilian-blockchain-startup-ecotrace-gets-funding-for-traceability-tech.

AIS. (2022). AIS redesigns the E-Waste ecosystem with blockchain on E-Waste+. For the first time in the Southeast Asia region, TGO and six green network partners have joined forces to power a sustainable environment push. Retrieved April 16, 2023, from Advanced Info Service website: https://sustainability.ais.co.th/en/update/news/279/

ais-redesigns-the-e-waste-ecosystem-with-blockchain-on-e-waste-for-the-first-time-in-the-southeast-asia-region-tgo-and-six-green-network-partners-have-joined-forces-to-power-a-sustainable-environment-push.

Bag, S., Viktorovich, D.A., Sahu, A.K. and Sahu, A.K. (2021). Barriers to adoption of blockchain technology in green supply chain management. Journal of Global Operations and Strategic Sourcing, 14(1): 104–133. https://doi.org/10.1108/JGOSS-06-2020-0027/FULL/XML.

Bagnoli, C. and Maura, A. (2021). Business Model Circolari. Torino: Giappichelli.

Bakarich, K.M., Castonguay, J. "Jack" and O'Brien, P.E. (2020). The use of blockchains to enhance sustainability reporting and assurance. Accounting Perspectives, 19(4): 389–412. https://doi.org/10.1111/1911-3838.12241.

Bigliardi, B., Campisi, D., Ferraro, G., Filippelli, S., Galati, F. and Petroni, A. (2020). The intention to purchase recycled products: Towards an integrative theoretical framework. Sustainability, 12(22): 9739. https://doi.org/10.3390/SU12229739.

Biloslavo, R., Bagnoli, C. and Edgar, D. (2018). An eco-critical perspective on business models: The value triangle as an approach to closing the sustainability gap. Journal of Cleaner Production, 174: 746–762. https://doi.org/10.1016/j.jclepro.2017.10.281.

Calandra, D., Secinaro, S., Massaro, M., Dal Mas, F. and Bagnoli, C. (2022). The link between sustainable business models and Blockchain: A multiple case study approach. Business Strategy and the Environment. https://doi.org/10.1002/BSE.3195.

Cision, P.R. Newswire. (2022). Hasiru Dala Innovations joins forces with Satma CE to give immutable waste traceability from collection to final product made from waste. Retrieved April 14, 2023, from https://www.prnewswire.com/in/news-releases/hasiru-dala-innovations-joins-forces-with-satma-ce-to-give-immutable-waste-traceability-from-collection-to-final-product-made-from-waste-836247047.html.

Clarke, V., Braun, V. and Hayfield, N. (2015). Qualitative psychology: A practical guide to research methods. Qualitative Psychology, 1–312. Retrieved from https://books.google.co.za/books?id=lv0aCAAAQBAJ.

Esmaeilian, B., Wang, B., Lewis, K., Duarte, F., Ratti, C. and Behdad, S. (2018). The future of waste management in smart and sustainable cities: A review and concept paper. Waste Management, 81: 177–195. https://doi.org/10.1016/J.WASMAN.2018.09.047.

Gatteschi, V., Lamberti, F., Demartini, C., Pranteda, C. and Santamaria, V. (2018). To blockchain or not to blockchain: That is the question. IT Professional, 20(2): 62–74. https://doi.org/10.1109/MITP.2018.021921652.

GlobeNewswire. (2022). Phantasma is Now a Carbon Negative Blockchain. Retrieved April 16, 2023, from GlobeNewswire website: https://www.globenewswire.com/en/news-release/2021/12/22/2356756/0/en/Phantasma-is-Now-a-Carbon-Negative-Blockchain.html.

Grushina, S.V. (2017). Collaboration by design: Stakeholder engagement in GRI sustainability reporting guidelines. Organization & Environment, 30(4): 366–385. Retrieved from https://www.jstor.org/stable/26408348.

Guo, S., Sun, X. and Lam, H.K.S. (2020). Applications of blockchain technology in sustainable fashion supply chains: Operational transparency and environmental efforts. IEEE Transactions on Engineering Management. https://doi.org/10.1109/TEM.2020.3034216.

Hsieh, H.-F. and Shannon, S.E. (2005). Three approaches to qualitative content analysis. Qualitative Health Research, 15(9): 1277–1288. https://doi.org/10.1177/1049732305276687.

Hua, W., Jiang, J., Sun, H. and Wu, J. (2020). A blockchain based peer-to-peer trading framework integrating energy and carbon markets. Applied Energy, 279: 115539. https://doi.org/10.1016/J.APENERGY.2020.115539.

Jabbour, C.J.C. (2013). Environmental training in organisations: From a literature review to a framework for future research. Resources, Conservation and Recycling, 74: 144–155. https://doi.org/10.1016/J.RESCONREC.2012.12.017.

Kewell, B., Adams, R. and Parry, G. (2017). Blockchain for good? Strategic Change, 26(5): 429–437. https://doi.org/10.1002/JSC.2143.

Klean Industries. (2022). Klean Industries Partners with RGH Systems to Develop a Waste Plastics to Energy Facility in Philippines. Retrieved April 16, 2023, from Klean Industries website: https://kleanindustries.com/investors/press-releases/partners-rgh-systems-develop-waste-plastics-to-energy-facility-philippines/.

Krippendorff, K. (2013). Content Analysis: An Introduction to its Methodology. Los Angeles.

Landrum, N.E. and Ohsowski, B. (2018). Identifying worldviews on corporate sustainability: A content analysis of corporate sustainability reports. Business Strategy and the Environment, 27(1): 128–151. https://doi.org/10.1002/bse.1989.

Massaro, M., Dumay, J., Garlatti, A. and Dal Mas, F. (2018). Practitioners' views on intellectual capital and sustainability: From a performance-based to a worth-based perspective. Journal of Intellectual Capital, 19(2): 367–386. https://doi.org/10.1108/JIC-02-2017-0033/FULL/XML.

Massaro, M., Secinaro, S., Dal Mas, F., Brescia, V. and Calandra, D. (2021). Industry 4.0 and circular economy: An exploratory analysis of academic and practitioners' perspectives. Business Strategy and the Environment, 30(2): 1213–1231. https://doi.org/10.1002/BSE.2680.

Miles, M., Huberman, A. and Saldaña, J. (2014). Sampling: Bounding the collection of data. Qualitative Data Analysis: A Methods Sourcebook, 26–30.

Mohsin, M., Naseem, S., Zia-ur-Rehman, M., Baig, S.A. and Salamat, S. (2023). The crypto-trade volume, GDP, energy use, and environmental degradation sustainability: An analysis of the top 20 crypto-trader countries. International Journal of Finance & Economics, 28(1): 651–667. https://doi.org/10.1002/IJFE.2442.

Motshweni, T. (2022). Blockchain tech gives waste pickers the opportunity to enter the formal economy – The Mail & Guardian. Retrieved April 14, 2023, from Mail & Guardian website: https://mg.co.za/thoughtleader/opinion/2022-09-12-blockchain-tech-gives-waste-pickers-the-opportunity-to-enter-the-formal-economy/.

Patelli, N. and Mandrioli, M. (2020). Blockchain technology and traceability in the agrifood industry. Journal of Food Science, 85(11): 3670–3678. https://doi.org/10.1111/1750-3841.15477.

Richter, A., Ng, K.T.W., Karimi, N. and Chang, W. (2021). Developing a novel proximity analysis approach for assessment of waste management cost efficiency in low population density regions. Sustainable Cities and Society, 65: 102583. https://doi.org/10.1016/J.SCS.2020.102583.

Romme, A. G. L., Avenier, M. J., Denyer, D., Hodgkinson, G. P., Pandza, K., Starkey, K., & Worren, N. (2015). Towards Common Ground and Trading Zones in Management Research and Practice. British Journal of Management, 26(3), 544–559. https://doi.org/10.1111/1467-8551.12110.

Ruzza, D., Dal Mas, F., Massaro, M. and Bagnoli, C. (2020). The role of blockchain for intellectual capital enhancement and business model innovation. Intellectual Capital in the Digital Economy, 256–265. Retrieved from https://www.taylorfrancis.com/chapters/edit/10.4324/9780429285882-24/role-blockchain-intellectual-capital-enhancement-business-model-innovation-daniel-ruzza-francesca-dal-mas-maurizio-massaro-carlo-bagnoli.

Sahoo, S., Kumar, S., Sivarajah, U., Lim, W.M., Westland, J.C. and Kumar, A. (2022). Blockchain for sustainable supply chain management: Trends and ways forward. Electronic Commerce Research, 2022: 1–56. https://doi.org/10.1007/S10660-022-09569-1.

Salah, K., Nizamuddin, N., Jayaraman, R. and Omar, M. (2019). Blockchain-based soybean traceability in agricultural supply chain. IEEE Access, 7: 73295–73305. https://doi.org/10.1109/ACCESS.2019.2918000.

Scekic, O., Nastic, S. and Dustdar, S. (2019). Blockchain-supported smart city platform for social value co-creation and exchange. IEEE Internet Computing, 23(1): 19–28. https://doi.org/10.1109/MIC.2018.2881518.

Secinaro, S., Dal Mas, F., Massaro, M. and Calandra, D. (2021). Exploring agricultural entrepreneurship and new technologies: Academic and practitioners' views. British Food Journal, ahead-of-print(ahead-of-print). https://doi.org/10.1108/BFJ-08-2021-0905.

Singh, A.K., Oguntoye, O. and Packard, L. (2022). Cleaning Up Crypto's Emissions: Why Policy Shouldn't Be Binary. London. Retrieved from https://www.institute.global/insights/tech-and-digitalisation/cleaning-cryptos-emissions-why-policy-shouldnt-be-binary.

Smith, A.E. and Humphreys, M.S. (2006). Evaluation of unsupervised semantic mapping of natural language with Leximancer concept mapping. Behavior Research Methods, 38(2): 262–279. https://doi.org/10.3758/BF03192778.

Smith, L., Carder, P., Bucy, T., Winfree, J., Brazier, J.F., Kaskie, B. and Thomas, K.S. (2021). Connecting policy to licensed assisted living communities, introducing health services regulatory analysis. Health Services Research, 56(3): 540–549. https://doi.org/10.1111/1475-6773.13616.

US EPA. (2021). EPA awards more than $230,000 for recycling and food waste prevention projects in Illinois, Michigan, and Ohio. Retrieved April 14, 2023, from United States Environmental Protection Agency website: https://www.epa.gov/newsreleases/epa-awards-more-230000-recycling-and-food-waste-prevention-projects-illinois-michigan.

World Economic Forum. (2023). 30 digital solutions to fuel industry decarbonization at scale. Retrieved April 16, 2023, from World Economic Forum website: https://www.weforum.org/agenda/2023/01/davos23-digital-solutions-to-power-decarbonization/.

Chapter 5

Improving Sustainable Supply Chains Using Blockchain Technology

A Review of the Literature

Albérico Rosário[1,*] and *Joana Carmo Dias*[2]

1. Introduction

Blockchain is classified as a disruptive technology based on advanced computing and innovation technologies characterized by decentralized integration of databases with open-source networking. Blockchain innovation is based on the idea that transactions and information can be recorded systematically through the cryptographic process in a public database (Schinckus, 2020). The decentralization of the network allows all stakeholders to contribute to the information validation since it involves a shared digital history and updating of records. The blockchain era emerged from the 2008 global financial crisis when people began questioning the operations of banks and financial institutions. The crisis highlighted the asymmetric information issue characterizing the relationship between investors and financial institutions and resulted in a widespread call for inclusion and transparency. As a result, the innovation's appeal is derived from the capability to support transparent data sharing, optimize business

[1] GOVCOPP, IADE – Universidade Europeia.
[2] COMEGI, Universidade Lusíada.
Email: joana.carmo.dias@universidadeeuropeia.pt
* Corresponding author: alberico@ua.pt

operations, improve efficiency in collaboration, and reduce operations costs (Bai et al., 2020). It aims at developing a system that differentiates itself from others, such as supply chains, by not necessarily incorporating trust in its functioning. Therefore, blockchain aims to alter the existing financial practices and develop new business systems and activities.

Given the rise in climate change, environmental, economic, and social issues, sustainability has become a topic of significant interest. Blockchain technology's capability to change industrial operations has led to its definition as a foundational technology that can be used to promote sustainability. For instance, its irrevocability and transparency grants all community members access to transaction and records and reduces the probability of altering or corrupting record or exchange processes (Schinckus, 2020). As a result, it has been identified as a potential innovation capable of achieving the 2030 sustainable goals developed by the United Nations Environment Programme (UNEP). For example, blockchain technology improves traceability of investments and ownerships, reducing fraud and corruption, misuse of funds, and potential data alterations (Lund et al., 2019). These capabilities can help achieve UN goals: goal 8 - good jobs and economic growth and goal 10 - reduced inequalities. It can also contribute to clean energy through enhanced climate finance flows and carbon emission trading. As a new technology, blockchain innovation can be exploited to achieve multiple sustainability goals and objectives.

In addition, blockchain technology can contribute to supply chain sustainability by influencing the supply chains' social, economic, and environmental performance. Social sustainability can include empowering trust by creating authority and trust in decentralized networks (Giungato et al., 2017). For instance, investors and other stakeholders have access to transactional information and trace finance flows and updates. The transparency and completeness of the information shared through these platforms create mutually trusting relationships among supply chain stakeholders (Rejeb and Rejeb, 2020). It can also be used for humanitarian causes, food safety, and social empowerment. For economic sustainability, blockchain technology provides value creation opportunities, operational efficiencies and market disintermediation. It also contributes to environmental sustainability by providing control and protection initiatives that carefully monitor production aspects such as energy consumption, emissions, and processing of raw materials (Giungato et al., 2017). Therefore, the purpose of this research is to examine the correlation between blockchain technology and sustainability and provide findings that can be used to exploit the sustainable opportunities created.

2. Methodological Approach

A Systematic Bibliometric Literature Review (LRSB) was carried out for data collection and analysis to provide insights on the various ways blockchain technology drives sustainability. The literature review is a systematic, comprehensive, and reproducible technique of collecting, assessing, and synthesizing data created by practitioners, researchers, and scholars (Okoli, 2015). Table 1 presents the screening methods adopted. For instance, to narrow down the search and ensure that only the most relevant documents were selected for synthesis, exact keywords such as "Sustainable Development", "Sustainable Supply Chains", "Environmental Sustainability", "Sustainable Performance", and "Social Sustainability" were used. The search was also limited to the subject area "Business, Management and Accounting" to narrow down the documents further. Of the identified articles, their relevance was first evaluated by screening title contents, followed by the abstract. If the researcher identified that the content matched the topic, a complete-article screening was conducted for inclusion. The database of scientific and/ or academic articles used was SCOPUS, the largest abstract and citation database of peer-review literature. However, we consider that the study has the limitation of considering only the SCOPUS database, excluding the other scientific and academic bases. The bibliographic search includes peer-reviewed scientific and/or academic documents published until September 2021.

Table 1. Screening methodology.

Database Scopus	Screening	Publications
Meta-search	keyword: Blockchain, Sustainability	458
First Inclusion Criterion	keyword: Blockchain, Sustainability Subject area Business, Management and Accounting	110
Second Inclusion Criterion	keyword: Blockchain, Sustainability Subject area Business, Management and Accounting Exact keywords: Sustainability, Sustainable Development, Sustainable Supply Chains, Environmental Sustainability, Sustainable Performance, Social Sustainability	61
Screening	keyword: Blockchain, Sustainability Subject area Business, Management and Accounting Exact keywords: Sustainability, Sustainable Development, Sustainable Supply Chains, Environmental Sustainability, Sustainable Performance, Social Sustainability Published until September 2021	

61 scientific and/or academic documents indexed in SCOPUS were analyzed in a narrative and bibliometric way to deepen the content themes that respond directly to the research question (Rosário, 2021; Raimundo and Rosário, 2021; Rosário et al., 2021; Rosário and Cruz, 2019). Of the 61 documents selected, 42 were articles; 7 conference papers; 7 reviews; 7 Book chapters; and 1 Editorial.

2.1 *Publication Distribution*

Figure 1 summarizes the published peer-reviewed literature for the 2018–2021 period. The publications were sorted as follows: Journal of Cleaner Production (10); International Journal of Information Management (5); International Journal of Production Research (5); Technological Forecasting and Social Change (4); British Food Journal, International Journal of Production Economics, Transforming Climate Finance and Green Investment with Blockchains, Transportation Research Part E Logistics And Transportation Review (2 each); the remaining publications have only one document each.

Between 2020-2021 there was great interest in Blockchain Sustainability publications.

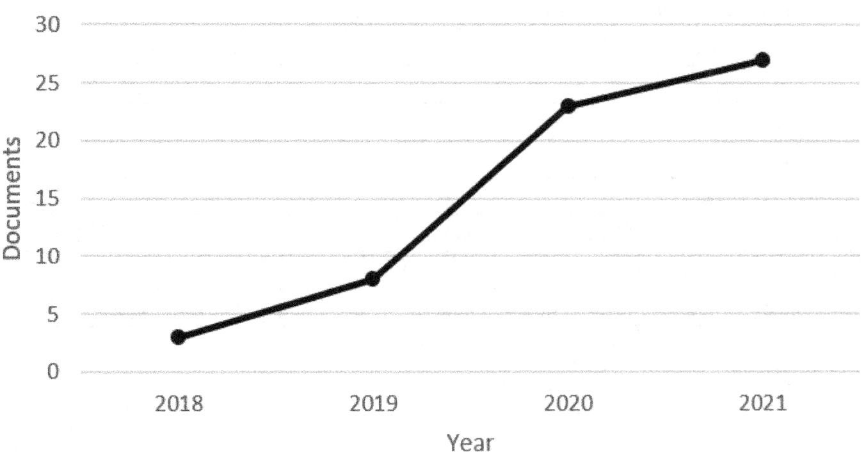

Figure 1. Documents by year.

In Table 2 we analyze per the Scimago Journal & Country Rank (SJR), the best quartile and h-index by publication. The International Journal of Information Management is the most quoted publication with 2,770(SJR), Q1 and h-index 114.

There is a total of 17 publications on Q1, 7 journals on Q2, 2 journals on Q3 and 1 journal on Q4. Journals from best quartile Q1 represent 47%

Table 2. Scimago journal & country rank impact factor.

Title	SJR	Best Quartile	H Index
International Journal of Information Management	2,770	Q1	114
Journal of Service Management	2,660	Q1	60
Journal of Business Logistics	2,610	Q1	79
International Journal of Production Economics	2,410	Q1	185
Technological Forecasting and Social Change	2,230	Q1	117
Business Strategy and the Environment	2,120	Q1	105
Transportation Research Part E Logistics and Transportation Review	2,040	Q1	110
Journal of Cleaner Production	1,940	Q1	200
International Journal of Production Research	1,910	Q1	142
Corporate Social Responsibility and Environmental Management	1,520	Q1	73
Futures	1,21	Q1	79
Marketing Letters	1,130	Q1	70
Industrial Management and Data Systems	0,990	Q1	103
Journal of Hospitality and Tourism Technology	0,970	Q1	31
Management Decision	0,920	Q1	98
Technology in Society	0,820	Q1	51
IEEE Transactions on Engineering Management	0,700	Q1	92
Benchmarking	0,640	Q2	61
British Food Journal	0,510	Q2	80
International Journal of Finance and Economics	0,510	Q2	39
Journal of Organizational Change Management	0,510	Q2	70
Journal of Sustainable Finance and Investment	0,450	Q2	16
International Journal of Productivity and Performance Management	0,420	Q2	61
Cogent Business and Management	0,350	Q2	16
Logforum	0,200	Q3	4
Developments in Corporate Governance and Responsibility	0,170	Q3	6
Strategic Direction	0,120	Q4	12
Proceedings of the European Conference on Innovation and Entrepreneurship Ecie	0,130	-*	6
Icbc 2019 IEEE International Conference on Blockchain and Cryptocurrency	0,340	-*	8

Table 2 contd. ...

...Table 2 contd..

Title	SJR	Best Quartile	H Index
International Conference on the European Energy Market Eem	0,270	-*	18
Entrepreneurship and Sustainability Issues	0	-*	25
Contributions to Management Science	0	-*	14
Transforming Climate Finance and Green Investment with Blockchains	-*	-*	-*
2020 IEEE Technology and Engineering Management Conference Temscon 2020	-*	-*	-*
2020 International Conference on Data Analytics for Business and Industry Way Towards a Sustainable Economy Icdabi 2020	-*	-*	-*
IEEE International Conference on Blockchain and Cryptocurrency Icbc 2021	-*	-*	-*
Technology Innovation Management Review	-*	-*	-*

Note: *data not available.

of the 61 journals titles; best quartile Q2 represents 19%, best quartile Q3 represents 6%, and best quartile Q4 represents 3% each of the titles. Finally 9 publications without data represent 25% of all publications.

As evident from Table 2, the majority of articles on blockchain sustainability is ranked in the Q1 best quartile index.

The subject areas covered by the 61 scientific articles were: Business, Management and Accounting (61); Engineering (25); Decision Sciences (22); Environmental Science (13); Computer Science (12); Energy (12); Social Sciences (11); Economics, Econometrics and Finance (8); Psychology (4); Agricultural and Biological Sciences (2).

In Figure 2 we can analyze the evolution of citations of articles published between 2018 and 2021. The number of quotes shows a positive net growth with an R2 of 46% for the period 2011–2021, with 2021 reaching 1041 citations.

The h-index was used to verify the productivity and impact of the published work, based on the largest number of articles included that had at least the same number of citations. Of the documents considered for the h-index, 16 were cited at least 16 times.

In Figure 3, we can find the results of a bibliometric study carried out to investigate and identify indicators on the dynamics and evolution of scientific information based on the main keywords. This study, through the use of the scientific software VOSviewer, aims to identify the main research keywords in blockchain sustainability studies. The linked

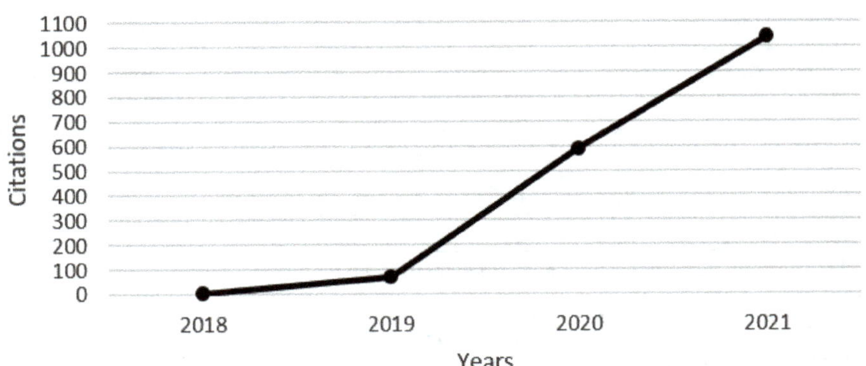

Figure 2. Evolution of citations between 2018 and 2021.

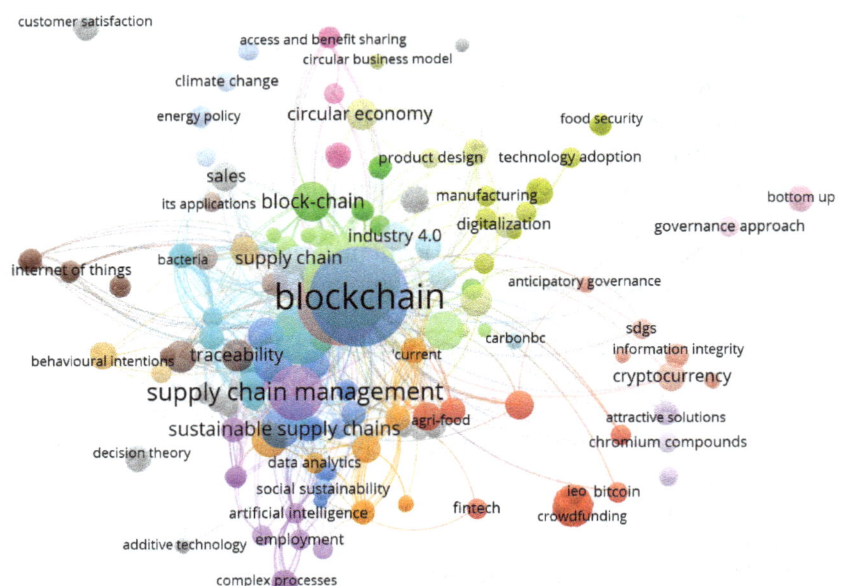

Figure 3. Network of all keywords.

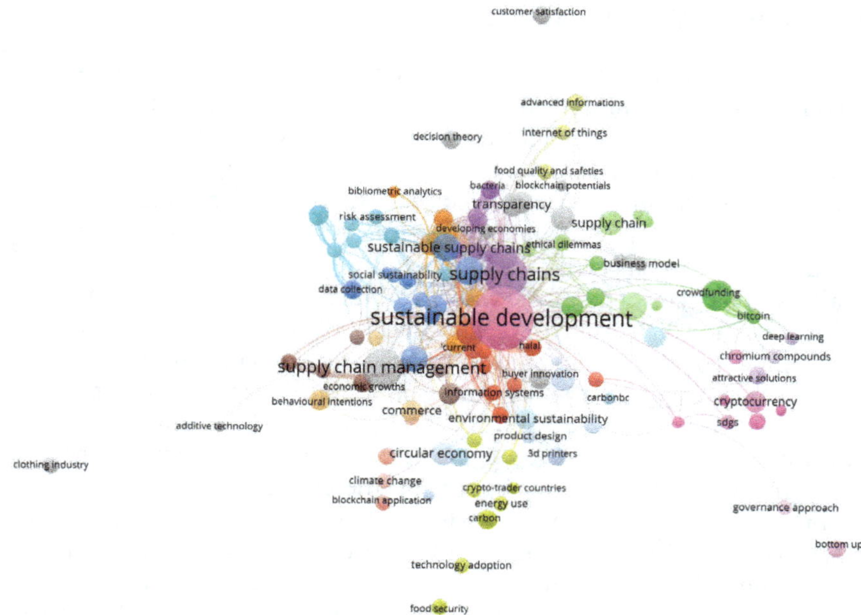

Figure 4. Network of linked keywords.

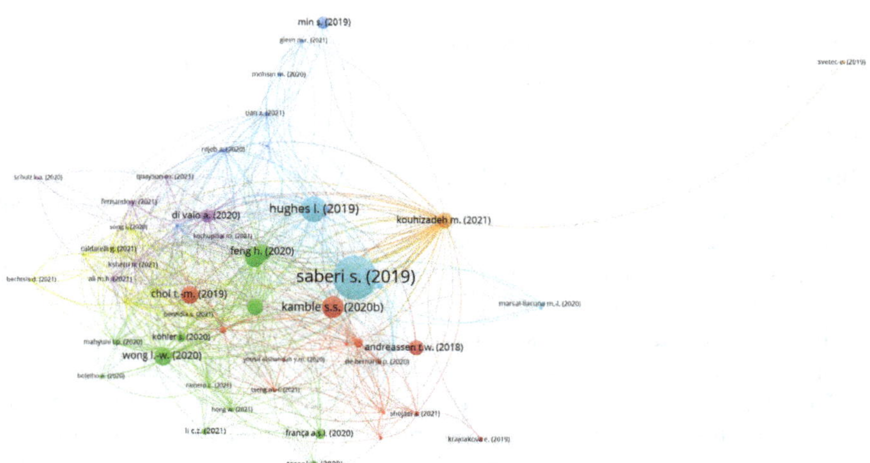

Figure 5. Network of bibliographic coupling.

keywords can be examined in Figure 4 making it possible to highlight the network of keywords that appear together/linked in each scientific article, allowing to know the themes studied by the researchers and to identify trends for future researches. In Figure 5, a profusion of bibliographic coupling with a unit of analysis of cited references is presented.

3. Theoretical Perspectives

As the world transitions to the digital age, emerging technologies have become catalysts in influencing current and future practices. For instance, decision-making and foresight are considerably dependent on big data analytics and artificial intelligence (AI). Other emerging technologies such as blockchain have attracted the interests of scientists, policy-makers, investors, and development practitioners. These technologies are expected to help achieve sustainability goals and confront societal challenges affecting people's wellbeing (Schulz et al., 2020). Various innovative 'FinTech' networks have been established to facilitate the development of blockchain innovations. Examples include the UN Climate Chain Coalition (CCC), the Stanford Center for Blockchain Research, and the European Union Blockchain Observatory and Forum. Blockchain innovations contribute to the expansion of digital infrastructures and the digitization of processes, increasing the efficiency and effectiveness of business and governance systems (Soni et al., 2021). Therefore, blockchain solutions are expected to contribute to economic, social, and environmental sustainability by providing advanced digital innovations. This section aims to explore how blockchain solutions contribute to these sustainability initiatives.

3.1 The Sustainable Management (SM) in Open Innovation Literature: Defining Key Concepts

Blockchain

Blockchain refers to a digital ledger of transactions that are encrypted in blocks of data chained together to create a chronological single-source of truth for the data. These blocks are then distributed and duplicated across a network of computer systems. Satoshi Nakamoto (pseudonym) introduced blockchain technology as an open distributed ledger tool to facilitate cryptocurrencies like bitcoin (Upadhyay et al., 2021). The bitcoin protocol was a secure, open, and transparent distributed ledger technology (DLT) that allowed partners to transact without trusted third parties. However, the complexity of the technology elicited negative criticism since most investors had a limited understanding of the technology's functioning (Tanev and Sandstrom, 2020). In recent years, developers and other interested parties have increased awareness of DLT's advantages, such as increased transparency and capability to reduce data manipulation and alterations, leading to increased research on its applications (Wong et al., 2020). As a result, blockchain has been classified as both a disruptive and foundational technology. Its attractive features include its use as a distributed ledger, increased transparency, decentralized control, complex technologies that prevent tamper or manipulations, permanent transactions, and open access (Upadhyay et al., 2021). In addition, blockchain technologies have developed into three versions, including Blockchain 1.0, 2.0, and 3.0, leading to diverse implementation influencing decisions to adopt blockchain innovations.

DLT runs on top of the Internet Protocol, recording transactions permanently and in a trustable way using distributed consensus algorithms and cryptographic techniques. Each user on the network has a private key to allow reading messages and public for use by others during encryption and must be connected to the point-to-point network (Upadhyay et al., 2021). The private key is used to sign a transaction after it is carried out before it is broadcasted. This process facilitates authentication and ensures that the information transmitted cannot be decrypted in case of an error (Rana et al., 2021). A consensus algorithm is used to order the transactions in timestamped blocks that create a block of chain and can be verified through automation and governance protocols to ensure it only contains valid transactions (Wong et al., 2020). The encryption mechanisms protect the data shared throughout the network from unauthorized access and build trust while eliminating third parties. Wong et al. (2020) identify two broad categories of blockchain: permission-less blockchains and permissioned blockchains. Under permission-less blockchain, the transactions are public, and users do not require permission to read, submit or engage in the consensus process (Rejeb and Rejeb, 2020). On the contrary, permissioned blockchains control users who can transact, and therefore, participation requires an invitation. The choice of the type of blockchain innovation adopted depends on the nature of transactions and the parties involved.

Sustainability

While industrialization and digitization have created opportunities to empower global populations, it has subjected the world to potential planetary disasters of unprecedented magnitude. For instance, there are increased concerns over how business practices contribute to climate change through carbon emissions (Saberi et al., 2019). As a result, sustainability has attracted significant attention from researchers, government and non-government agencies, activists, and the general public. However, despite the increased use of the terms "sustainability" and "sustainable development," Moore et al. (2017) indicate that they lack comprehensive definitions, affecting the capability to operationalize and measure. For instance, Robertson (2017) defines sustainability as systems and processes that can persist and operate for an extended period on their own. Scoones (2016) defines it as the capability to meet present needs without compromising the future. In addition, the author indicates that sustainability can be defined as a system's capability to bounce back from stresses and shocks. Despite the variations in wording, the definitions identify various elements of sustainability, including the capacity to endure, sustain, and maintain.

Sustainability has three dimensions: economic, environmental, and social. Economic sustainability involves practices that facilitate long-term economic growth without undermining other aspects of society, such as cultural, social, and environmental aspects (Li et al., 2021). It indicates the need to use, protect, and sustain human and material resources to create long-term sustainable values through practices such as recycling and optimal use. Social sustainability identifies the effects of economic activities on people (Moore et al., 2017). For example, the concept emphasizes improving the quality of the organization's relationships and engagement with its stakeholders (Wong et al., 2020). Businesses directly or indirectly impact workers, communities, customers, and supply chains, thus potentially influencing issues such as health, education, and human rights (Shojaei et al., 2021). Environmental sustainability refers to protecting and conserving natural resources and the global ecosystem to ensure current and future health and well-being. The social, economic, and environmental sustainability dimensions are interrelated and can significantly affect each other (Robertson, 2017). For example, economic development can have negative implications on the environment. Sustainability is based on the need to ensure that current practices do not compromise the capability of future generations in meeting their needs.

Economic Sustainability

Leveraging blockchain technologies can generate sustainability-induced changes. For instance, it can contribute to human and environmental sustainability and social values. Some aspects of blockchain-supported economic sustainability include market disintermediation, value creation opportunities, and operational efficiencies.

Market Disintermediation

Few intermediaries monopolize the flow of products and materials in the current market, resulting in higher costs, complex systems, and customer rejections. With disintermediation, intermediaries between customers and producers during transactions are reduced (Song et al., 2020). Blockchain technologies contribute to this opportunity by directly connecting buyers and sellers through an interconnected network where they transact and update records accordingly (Bechtsis et al., 2021). The innovations serve as a central authority whose primary focus is to validate transactions. Trading partners can combine blockchain transactions and smart contracts to eliminate the need for trusted third parties in transshipment operations.

Value Creation Opportunities

Blockchain technologies' monitoring and control capabilities can increase the visibility of underutilized resources such as machinery, warehouse, and equipment. As a result, supply chain partners within a shared model can allocate resources more efficiently and effectively (Svetec et al., 2019). In addition, blockchain innovations increase an organization's digital capabilities of creating decentralized and collaborative logistics that match the demand faced by companies; they increase utilization and availability of supply chain logistics resources, and assemble many supply chain actors (Andreassen et al., 2018). The strict monitoring and scrutiny of transactions and business processes under blockchain technologies can lead to the creation of fair economic models based on adequate data sharing (Tseng et al., 2021). This is unlike in the current financial environment, which is characterized by asymmetric information flow, where banks and financial institutions limit the amount of information accessible to investors (Chenli et a., 2019). With blockchain's transparency strategy, the investors will be more included in the transaction processes. The inclusivity and collaboration will significantly contribute to value creation by providing a new environment for entrepreneurship and developing new companies and business ventures.

Additionally, the application of blockchain technologies enables companies to benefit from economic incentives regardless of their size. For instance, blockchain simplifies crowdfunding and reduces costs and entry barriers (França et al., 2020). In this case, small and SMEs can attract investments and cheap capital from all over the world, generating competitive resources and creating growth and wealth (Felipe Munoz et al., 2021). Consequently, blockchain solutions can achieve sustainable goals of reducing inequalities through wealth distribution and equal access to resources. This flow of investments will enable companies to create new values and meet the expectations of their customers, thus reducing the possibilities of being outperformed by their competitors.

Operational Efficiencies

Businesses can adopt blockchain innovations to streamline their operations and increase the supply chain's responsiveness and efficiency. They can benefit from automation, traceability, authentication, and verification efficiencies facilitating multiparty transactions (Tham and Sigala, 2020). For instance, blockchain technology enables end-to-end information sharing and product tracking, increasing transparency and visibility that can be used to reduce non-value-adding practices that contribute to inefficiencies (Gleim and Stevens, 2021). The traceability capability allows companies to track products' progress and movements across the supply chain and the inventory level to enable effective decision-making. Besides, it eliminates inefficiencies associated with complex bureaucratic processes, archaic procedures, strict institutional conditions, and trade-related paperwork.

3.2 Social Sustainability

Blockchain contributes to social welfare by providing a trusted computing platform that emphasizes transparency, trust, and inclusivity. For instance, adopting blockchain solutions creates mutually trusting relationships among trading partners (Hughes et al., 2019). Blockchain-supported social sustainability can be illustrated from the perspectives of support for humanitarian causes, increased food safety, and empowering trust.

Support for Humanitarian Causes

Business organizations adopt emerging technologies to aid their social responsibilities and humanitarian supply chain functions. For example, blockchain solutions can be used to eliminate delays caused by political barriers, bureaucracy, and paperwork during crises by streamlining financial assistance (Kamble et al., 2020). With its advanced technologies, blockchain technology allows users to access and share adequate information for decision-making, beneficiaries access the required goods and services with minimal delays, and investors can control the accuracy, quality, and quantity of critically needed items during disaster management (Upadhyay et al., 2021). Humanitarian causes associated with critical emergencies can use the technology to build solidarity among people, reinforce trust, promote mutual communications, and trace donations to enhance transparency and ensure victims get the funds and items sent to them (Kochupillai et al., 2021). In addition, transparency can counteract corruption, abuse of human rights, and child labor. It can also support the financial integration of smallholding businesses, farmers, and unbanked populations (Tian et al., 2021). Organizations and individuals involved in humanitarian causes and social empowerment should optimize the benefits of blockchain technology to achieve social goals.

Increased Food Safety

Adopting blockchain innovations can help eliminate the growing health concerns over food safety. Blockchain-enabled traceability increases accountability, transparency, and efficiency that can influence the public perception of the food industry (Hong et al., 2020). For instance, food manufacturers and retailers can quickly respond to food safety issues due to the rapid information transmission, reducing the probability of spreading foodborne diseases (Botelho et al., 2020). In addition, end-to-end information sharing and product sharing can allow secure tracing of food-related data entries to ensure consumers access food in time before the expiration dates (Feng et al., 2020). The traceability can also enable fast contamination identification during a crisis to reduce the number of affected people and minimize mass anxiety that can further affect global societies and undermine local and international political landscape and economic progress.

Empowering Trust

Trust has become a significant challenge affecting online transactions due to an increase in unreliable businesses and scammers. This problem undermines the growth of e-commerce since internet users and organizations are concerned about the safety and security of their data and the authenticity of online sales (Köhler and Pizzol, 2020). Blockchain innovations can help eliminate these challenges by providing trading parties to trace and monitor product progress. Blockchain infrastructure and systems build trust by directly connecting buyers and sellers through transparent transactions and processes (Tiwari et al., 2019). For instance, all community members can access transactions and records, which cannot be altered or corrupted. Combining the transparency and irrevocability characteristics helps build trust and aligns with United Nations Environment Programme (UNEP) goals 1 (no poverty), 10 (reduce inequalities), and 11 (sustainable cities and communities) that emphasize empowering communities (Schinckus, 2020, p. 3). However, achieving these goals requires public engagement in social, economic, and environmental practices. Blockchain technology contributes to these goals by providing an inclusive platform where people and organizations, regardless of size and status in society, can access complete information and participate in transactions (Krajnakova et al., 2019). The platform's conditions create an environment necessary in building mutually trusting relationships that promote engagement and collaboration. In addition, it creates an atmosphere of honesty, fairness, and ethical behavior that facilitates information and resource allocation and sharing.

3.3 *Environmental Sustainability*

Environmental sustainability has become a significant focus area given the increased concerns over environmental degradation and climate change.

Blockchain technologies can be used to address environmental concerns by reducing the ecological footprint, a measure of the environmental impact of human activities in terms of the natural resources and ecosystem services required to support them (Fernando et al., 2021). It provides technologies that monitor energy production activities, energy consumption, raw materials processing, and gas emissions (Tozanlı et al., 2020). The information gathered throughout these processes can be used to implement environmental protection and control measures (Harris, 2018). Consequently, the transacting partners can be actively engaged in low-carbon energy initiatives and other environmental protection programs.

Blockchain can contribute to environmental sustainability by fostering environmental cooperation between firms and supply chain partners. For instance, companies can access real-time data from production processes that facilitate precise scheduling and allocation of resources (Krajnakova et al., 2019). This opportunity can help reduce resource overstretch that often leads to exploitation and involvement in environmentally unfriendly operations. For example, collaborations within the network can help reduce unnecessary transportation processes that contribute to gas emissions. In addition, the firms can trace carbon footprints and effectively trade carbon assets in the green-asset market. Therefore, optimizing blockchain solutions can help achieve sustainable clean energy goals, protecting the planet, and safeguarding life on earth and below water.

3.4 *Improving Sustainable Supply Chains Using Blockchain Technology*

A supply chain involves various points of goods manufacture and distribution from procurement to dispatch to consumers. The involvement of various locations and stations throughout the supply chain lines and lack of transparency makes it difficult to track progress and to sustain the value of goods and services(Mahyuni et al., 2020). Besides, the complexity of the value chain networks makes it challenging to identify those responsible for illegal activities, poor working conditions, forced labor, violation of human rights, and counterfeiting (Mangla et al., 2021). Blockchain technologies provide potential solutions that can solve these problems, such as a decentralized structure that allows multiple-party participation and increased access to information (Yousif Alsharidah and Alazzawi, 2020). Kshetri (2021) identifies this decentralization alongside two other features, immutability and cryptography-based authentication, as the primary blockchain characteristics associated with sustainability.

Decentralization

Blockchain technology embeds its value proposition in decentralization. For instance, blockchain eliminates the need for trusted third parties when conducting transactions through decentralized models, thus reducing costs and increasing efficiency (Kshetri, 2021). In addition, the high transparency enhances the technology's trust capabilities, building optimism on its capability to influence multiple industrial changes.

Immutability

Immutability indicates that the processes cannot be changed. Immutable in programming suggest that an object created and recorded in a software tool cannot be altered. According to Kshetri (2021), the immutability of blockchain technologies ensures that transactions are indelible, auditable, and cannot be forged. Eliminating data manipulation ensures that regulators and policy-makers have access to accurate information for decision-making and policy implementation on sustainability issues.

Cryptography-based Authentication

Cryptography-based authentication ensures that the information is only accessible to authorized persons. For example, participants in a transaction need to use cryptography-based digital signatures to verify their identities and acquire appropriate access to specific networks and records (Di Vaio and Varriale, 2020). This process involves using a private key, a long and random alphanumeric code, to sign transactions (Kshetri, 2021). A public key can be generated from the private key to facilitate information sharing and enable measuring and tracking sustainability-based outcomes.

Moreover, blockchain offers transparency, real-time information updates, and manipulation resistance capabilities to reduce supply chain risks and problems. These opportunities can be used to address challenges faced throughout traditional supply chain points, such as product mishandling, delayed delivery, theft, and fraud (Ali et al., 2021). Jayawardhana and Colombage (2020) explain that these solutions are achieved through blockchain's primary pillar; governance and responsibility. While governance involves regulations and control, responsibility involves increased accountability. For instance, tracing carbon footprints can raise awareness of environmental sustainability among supply chain partners, leading to responsible business behaviors (Marsal-Llacuna, 2020). In addition, the tracking and monitoring of products and information disseminated can lead to enhanced product quality, lowered operational costs, and reduced fraud. These opportunities illustrate blockchain's potential contribution to effective and efficient supply chains.

Blockchain technology can be used to enhance the new product development (NPD) process. According to Benzidia et al. (2021), the NPD process requires collaborative efforts from suppliers and compliance with sustainability practices to guarantee access to substantial information such as digital design. Blockchain provides a networked platform that allows suppliers to participate in information sharing and other appropriate transactions. Besides, as the world transitions into Industry 4.0, NPD has become significantly reliant on data. Thus, information transfer has become a critical element of smart manufacturing dependent on big data technologies and AI-based innovations (Min et al., 2019). Decision-making on smart products designs, features, and benefits are based on data analysis and interpretation. However, the seamless information transfer between the stakeholders is characterized by traceability, security, and reliability challenges, affecting the supply network. Blockchain technology provides solutions needed to overcome these challenges and contribute to sustainable Industry 4.0 growth.

Blockchain as a disruptive technology can enhance sustainable supply chain management (SSCM) through close monitoring of product flows within supply chains. In addition, tracking environmental and social threats that undermine the health and safety of others, boosts confidence in a process that traces the authenticity (Kouhizadeh et al., 2021) of sustainable products. These can be achieved through the private/permissioned or public/permissionless dimensions of blockchain technology. Under the public blockchain technology, any user can access the information and records and join the network, while under the private blockchain, only those with authorized access can participate in the transactions (Mishra and Kaushik, 2021). Regardless of the dimension adopted, blockchain solutions ensure that appropriate parties access real-time data on the records and transactions. This capability contributes to close monitoring of product flows and timely identification of information and transactions that threaten the environment, economy, and social welfare.

Blockchain can be used to facilitate the development of sustainable supply chains in developing countries. The developing world is characterized by low levels of trust and high intermediation costs, which affect the capability to address sustainability issues (Kshetri, 2021). Blockchain facilitates disintermediation, builds trust, and decentralizes markets, providing potential tools for addressing the problems in developing countries. It also eliminates personal and institutional intermediation, which often contributes to fraud and corruption affecting sustainable supply chains in these countries (Mohsin et al., 2020). In addition, organizations are under increasing pressure from regulators, activists, and consumers to adopt sustainable supply chains (Choi and Luo, 2019). For instance, the European Union recommended integrating blockchain technologies to increase supply chain visibility in developing countries. As a significant trading partner, the EU can influence the adoption of these emerging technologies to improve companies' ethical behaviors throughout the supply chain networks. A Forbes Africa survey reported that 66% of participants, where 73% of these respondents were millennials, would pay for ethically and sustainably sourced products (Kshetri, 2021). These statistics reflect the increased demand to adopt sustainable business practices to acquire a competitive advantage in developing countries. Blockchain technology adoption would allow companies to market their products and services as sustainable by increasing supply chain visibility and access to community members' information.

Despite these developments in the supply chain, blockchain is associated with multiple barriers that undermine practical application. Implementing blockchain innovations in supply chains faces technological (T), organizational (O), and environmental (E) barriers (Kouhizadeh et al., 2021), that is, challenges along all dimensions of the TOE theoretical framework of technology adoption. As an emerging technology, barriers can vary from resistance based on lack of information on the technology to inadequate resources to acquire the technologies.

3.5 *Technological Barriers*

When deciding to adopt a technology, users will consider its technical capability, availability, difficulty, and complexity. According to Kouhizadeh et al. (2021), blockchain's primary limitation is immaturity as a technology, affecting its interoperability, scalability, and usability. These technical issues are associated with higher latency and lower throughput rates, limiting the volume of transactions recorded within a specific period (Caldarelli et al., 2021). Therefore, adopting blockchain technologies in the supply chain might not help firms achieve the desired optimal objectives since the infrastructures need further development.

Although blockchain technology has been introduced as a secure innovation that cannot be hacked and crashed, security concerns have been rampant. For example, system attacks and hacks in the cryptocurrency environment have raised concerns over the vulnerability of blockchain technology (Kouhizadeh et al., 2021). Another concern is the accessibility of IT infrastructure to supply chain partners, leading to the unintentional exclusion of some key players. While the data immutability strengthens authenticity and reliability, it can be a challenge in case of errors and inaccurate reports (Caldarelli et al., 2021). Immutability means that the data within the blockchain cannot be changed. Therefore, when errors in information shared are identified, participants cannot change the records, meaning that the inaccuracies will be forever within the blockchain network.

3.6 Organizational Barriers

Acquiring and maintaining blockchain technology requires appropriate hardware and software. The additional costs undermine organizational commitment to embrace the new technology, especially when the potential benefits cannot be quantified. New technologies are often expensive for the firm due to the demand for new talent with the required knowledge and skills as well as process infrastructures (Choi and Luo, 2019). Therefore, the need for additional investments lead to a low adoption rate. Besides, the risks of implementing an emerging technology influences top or middle management commitment. Leadership support is critical in adopting and implementing technologies since leaders are involved in the approval of human and physical resources, planning, and actualization of strategies (Niu et al., 2021). The management of risk-averse companies may not perceive blockchain innovation as a core element of their goals and mission.

Lack of adequate information on blockchain innovation is a significant adoption barrier. As an immature technology, the public understanding of blockchain technology is low and causes resistance. Since sustainability is a relatively new organizational practice, adopting an emerging technology to aid its implementation can influence perceived ease of use (Choi and Luo, 2019). In addition, the power imbalance within the supply chain networks affects participants' access to information and IT infrastructure (Parmentola et al., 2021). For example, large corporations have adequate resources to acquire appropriate infrastructure and training workforce while small and SMEs have limited financial capacity. Thus, the technology adoption and implementation rate within the supply chain network will differ, affecting efficiency and effectiveness.

3.7 Environmental Barriers

The environmental context in the supply chain can be demonstrated in two categories; the inter-organizational, and the broader external environment. Some issues from the inter-organizational view include integrating sustainable and technical practices into the supply chain, data protection and privacy within the inter-organizational systems, and geographical and cultural differences among supply chain partners (Kamble et al., 2020). In some cases, supply chain partners may have varying goals and priorities, leading to ineffective communication and collaboration. The differences can undermine efforts to integrate sustainability goals and practices and blockchain technology capabilities. In addition, some collaborating firms may lack the sustainability knowledge needed to adopt sustainable practices across the supply chain (Poberezhna, 2018). Thus, embracing the new blockchain technology only adds to the confusion and complexity of the practices.

Organizations perceive their information as a competitive edge and are reluctant to share with other supply chain partners. Therefore, the privacy and confidentiality concerns within the inter-organizational systems reinforce their resistance to information sharing. However, Kouhizadeh et al. (2021) indicate that firms are willing to share data if it creates value for their clients and guarantees that the proprietary information will be confidential. However, the regulatory issues in blockchain technology provide minimal information-sharing policies, indicating a threatening policy gap that may further increase data privacy concerns (Kamble et al., 2020). Some sensitive sustainability information can pose potential legal and ethical concerns, affecting organizations' willingness to share. In addition, cultural and geographical differences can affect blockchain implementation in the supply chain since partners can adopt different performance tools and systems (Kouhizadeh et al., 2021). These variations may specifically affect social sustainability due to the heterogeneity of global cultures and their impacts on definitions and understanding of sustainability. Thus, while blockchain innovations provide multiple opportunities and benefits for the supply chains, these barriers may undermine adoption and integration into existing systems.

4. Blockchain as an Innovation for New Sustainable Business Model

A business model (BM) indicates an organization's logic to create, deliver, and capture value. Managers use BM as a conceptual tool to make strategic decisions and guide implementation processes by taking a system-level holistic approach to demonstrate how the company does business (Tiscini et al., 2020). A sustainable business model (SBM) can be defined as a business model that incorporates sustainability, principles, goals, and concepts to facilitate value creation (Massaro et al., 2020). It is used to explain how a company operationalizes its business strategy based on governance, strategy and management, systems, productions, and operations. SBMs improve environmental, social, and economic performance through innovative solutions that transform material and resources into products and services with economic and ecological value.

Despite the increased awareness of the significance of sustainability as a core component of business success, companies are struggling to integrate the concept into their BMs. Tiscini et al. (2020) explain that firms in the current business environment show unsustainable nature of production and consumption, indicating the need for new BMs, greater trust, and higher stakeholder engagements. Blockchain opportunities provide these solutions through advanced infrastructures that build trust, promote transparency, and allow participants access to information and records. Massaro et al. (2020) indicate that blockchain can simplify supply chains by reducing risks in trade finance, promoting inclusive economic systems, generating smarter market data, and increasing access to financial services. Thus, optimizing blockchain innovations can help firms overcome the various challenges experienced throughout the supply chain network ranging from trust to stakeholder involvement.

Blockchain uses a Distributed Ledger Technology (DLT) system, which differs from the traditionally used centralized ledger systems for data storage and management. DLT allows independent parties' hardware to document and synchronize publicly announced transactions to ensure that all parties have access to appropriate information (De Bernardi et al., 2020). This inclusivity and transparency contribute to the value co-creation process in SBMs. It helps bypass power imbalances that limit information and participation of supply chain partners based on their influence since all the parties within the network have access to unchangeable data (Rainero and Modarelli, 2021). For instance, Massaro et al. (2020) indicate that the blockchain distributed system requires all members to agree on the rules through a "census protocol," meaning equal participant engagement. As a valid SBM innovation, blockchain facilitates efficient collaboration through huge benefits such as data transparency and integrity (Quayson et al., 2021). In addition, given the increased consumer demands for sustainability, organizational decision-making has significantly become reliant on data technologies. Thus, information sharing through blockchain can contribute to innovative business models responsive to current consumer needs and expectations.

A circular Economy (CE) is a critical component of sustainability associated with sustainable production and manufacturing. According to Bressanelli et al. (2021), CE is an economic system that takes a regenerative approach to handle global problems such as waste management, climate change, biodiversity loss, and pollution. It emphasizes creating a closed-loop system that reduces the use of resource inputs, eliminates waste production and carbon emissions through processes such as reusing, sharing, repairing, refurbishing, recycling, and remanufacturing (Huynh, 2021). Bressanelli et al. (2021) identify blockchain as a CE enabler that can facilitate access to data needed for decision-making and product management throughout its entire life cycle. Blockchain technology can contribute to CE by easing the communication between stakeholders such as manufacturers and consumers to eliminate packaging waste.

Blockchain provides real-time data on product flows and market changes that manufacturers can use to establish sustainable practices for effective CE.

5. Conclusions

This study aims to show blockchain technology as an innovation that can be integrated into sustainability practices to develop new digital solutions that respond to current challenges. After a systematic search of the literature and analysis of the results we found that blockchain-supported technologies can lead to economic sustainability by providing value creation opportunities, increasing operational efficiencies, reducing costs, and promoting market disintermediation. In addition, blockchain can contribute to social sustainability by enhancing food safety through strict control and monitoring, supporting humanitarian causes, and ensuring the social empowerment of different communities. It can further be used to achieve environmental sustainability through the use of its control and monitoring features that trace production aspects such as gas emissions and energy consumption. Thus, adopting the appropriate blockchain technologies can drive sustainability.

However, blockchain integration into supply chains faces multiple challenges. For instance, some companies fear that transparent data sharing within the network threatens their competitive edge and can affect data privacy and confidentiality. Although the innovation has been introduced as secure, system attacks and hacks have significantly affected users' perceptions of its safety. Another adoption barrier has been the immaturity of the technology, which is associated with inadequate information and understanding of its operations and benefits. The need to purchase and maintain hardware and software, and retain qualified programmers results to additional costs that affect managers and investors' commitment to its implementation. However, since blockchain is a relatively new technology, the progress in development and increased awareness of its opportunities is expected to reduce resistance and increase the adoption rate.

Acknowledgements

We would like to express our gratitude to the Editor and the Referees. They offered extremely valuable suggestions or improvements. The authors were supported by the GOVCOPP Research Center of Universidade de Aveiro and COMEGI from Universidade Lusíada.

References

Ali, M.H., Chung, L., Kumar, A., Zailani, S. and Tan, K.H. (2021). A sustainable blockchain framework for the halal food supply chain: Lessons from Malaysia. Technological Forecasting and Social Change, 170. doi:10.1016/j.techfore.2021.120870.

Andreassen, T.W., Lervik-Olsen, L., Snyder, H., Van Riel, A.C.R., Sweeney, J.C. and Van Vaerenbergh, Y. (2018). Business model innovation and value-creation: The triadic way. Journal of Service Management, 29(5): 883–906. doi:10.1108/JOSM-05-2018-0125.

Bai, C. and Sarkis, J. (2020). A supply chain transparency and sustainability technology appraisal model for blockchain technology. International Journal of Production Research, 58(7): 2142–2162. doi:10.1080/00207543.2019.1708989.

Bai, C.A., Cordeiro, J. and Sarkis, J. (2020). Blockchain technology: Business, strategy, the environment, and sustainability. Bus. Strategy Environ., 29(1): 321–322.

Bechtsis, D., Tsolakis, N., Iakovou, E. and Vlachos, D. (2021). Data-driven secure, resilient and sustainable supply chains: Gaps, opportunities, and a new generalized data sharing and data monetisation framework. International Journal of Production Research. doi:10.1080/00207543.2021.1957506.

Benzidia, S., Makaoui, N. and Subramanian, N. (2021). Impact of ambidexterity of blockchain technology and social factors on new product development: A supply chain and industry 4.0 perspective. Technological Forecasting and Social Change, 169. doi:10.1016/j.techfore.2021.120819.

Botelho, A., Silva, I.R., Ribeiro, L., Lopes, M.S. and Au-Yong-Oliveira, M. (2020). Improving food transparency through innovation and blockchain technology. Paper Presented at the Proceedings of the European Conference on Innovation and Entrepreneurship, ECIE, 2020-September, 128–136. doi:10.34190/EIE.20.034.

Bressanelli, G., Pigosso, D.C.A., Saccani, N. and Perona, M. (2021). Enablers, levers, and benefits of circular economy in the electrical and electronic equipment supply chain: A literature review. Journal of Cleaner Production, 298. doi:10.1016/j.jclepro.2021.126819.

Caldarelli, G., Zardini, A. and Rossignoli, C. (2021). Blockchain adoption in the fashion sustainable supply chain: Pragmatically addressing barriers. Journal of Organizational Change Management, 34(2): 507–524. doi:10.1108/JOCM-09-2020-0299.

Chenli, C., Li, B., Shi, Y. and Jung, T. (2019). Energy-recycling blockchain with proof-of-deep learning. Paper presented at the ICBC 2019—IEEE International Conference on Blockchain and Cryptocurrency, 19–23. doi:10.1109/BLOC.2019.8751419.

Choi, T. and Luo, S. (2019). Data quality challenges for sustainable fashion supply chain operations in emerging markets: Roles of blockchain, government sponsors and environment taxes. Transportation Research Part E: Logistics and Transportation Review, 131: 139–152. doi:10.1016/j.tre.2019.09.019.

De Bernardi, P., Azucar, D., Forliano, C. and Franco, M. (2020). Innovative and sustainable food business models. Springer. doi:10.1007/978-3-030-33502-1_7.

Di Vaio, A. and Varriale, L. (2020). Blockchain technology in supply chain management for sustainable performance: Evidence from the airport industry. International Journal of Information Management, 52. doi:10.1016/j.ijinfomgt.2019.09.010.

Felipe Munoz, M., Zhang, K., Shahzad, A. and Ouhimmou, M. (2021). LogLog: A blockchain solution for tracking and certifying wood volumes. Paper presented at the IEEE International Conference on Blockchain and Cryptocurrency, ICBC 2021, doi:10.1109/ICBC51069.2021.9461153.

Feng, H., Wang, X., Duan, Y., Zhang, J. and Zhang, X. (2020). Applying blockchain technology to improve agri-food traceability: A review of development methods, benefits, and challenges. Journal of Cleaner Production, 260. doi:10.1016/j.jclepro.2020.121031.

Fernando, Y., Rozuar, N.H.M. and Mergeresa, F. (2021). The blockchain-enabled technology and carbon performance: Insights from early adopters. Technology in Society, 64. doi:10.1016/j.techsoc.2020.101507.

França, A.S.L., Amato Neto, J., Gonçalves, R.F. and Almeida, C.M.V.B. (2020). Proposing the use of blockchain to improve the solid waste management in small municipalities. Journal of Cleaner Production, 244. doi:10.1016/j.jclepro.2019.118529.

Giungato, P., Rana, R., Tarabella, A. and Tricase, C. (2017). Current trends in the sustainability of bitcoins and related blockchain technology. Sustainability, 9(12): 2214.

Gleim, M.R. and Stevens, J.L. (2021). Blockchain: A game-changer for marketers? Marketing Letters, 32(1): 123–128. doi:10.1007/s11002-021-09557-9.

Harris, A. (2018). A conversation with masterminds in blockchain and climate change. Transforming Climate Finance and Green Investment with Blockchains, pp. 15–22. doi:10.1016/B978-0-12-814447-3.00002-1.

Hong, W., Mao, J., Wu, L. and Pu, X. (2021). Public cognition of the application of blockchain in food safety management—Data from china's zhihu platform. Journal of Cleaner Production, 303. doi:10.1016/j.jclepro.2021.127044.

Hughes, L., Dwivedi, Y.K., Misra, S.K., Rana, N.P., Raghavan, V. and Akella, V. (2019). Blockchain research, practice, and policy: Applications, benefits, limitations, emerging research themes, and research agenda. International Journal of Information Management, 49: 114–129. doi:10.1016/j.ijinfomgt.2019.02.005.

Huynh, P.H. (2021). Enabling circular business models in the fashion industry: The role of digital innovation. International Journal of Productivity and Performance Management, doi:10.1108/IJPPM-12-2020-0683.

Jayawardhana, A. and Colombage, S. (2020). Does blockchain technology drive sustainability? An exploratory review. Governance and Sustainability. doi:10.1108/S2043-052320200000015002.

Kamble, S.S., Gunasekaran, A. and Gawankar, S.A. (2020). Achieving sustainable performance in a data-driven agriculture supply chain: A review for research and applications. International Journal of Production Economics, 219: 179–194. doi:10.1016/j.ijpe.2019.05.022.

Kamble, S.S., Gunasekaran, A. and Sharma, R. (2020). Modeling the blockchain-enabled traceability in the agriculture supply chain. International Journal of Information Management, 52. doi:10.1016/j.ijinfomgt.2019.05.023.

Kochupillai, M., Gallersdörfer, U., Köninger, J. and Beck, R. (2021). Incentivizing research and innovation with agrobiodiversity conserved *in situ*: Possibilities and limitations of a blockchain-based solution. Journal of Cleaner Production, 309. doi:10.1016/j.jclepro.2021.127155.

Köhler, S. and Pizzol, M. (2020). Technology assessment of blockchain-based technologies in the food supply chain. Journal of Cleaner Production, 269. doi:10.1016/j.jclepro.2020.122193.

Kouhizadeh, M., Saberi, S. and Sarkis, J. (2021). Blockchain technology and the sustainable supply chain: Theoretically exploring adoption barriers. International Journal of Production Economics, 231. doi:10.1016/j.ijpe.2020.107831.

Krajnakova, E., Svazas, M. and Navickas, V. (2019). Biomass blockchain as a factor of energetical sustainability development. Entrepreneurship and Sustainability Issues, (63): 1456–1467. doi:10.9770/jesi.2019.6.3(28).

Kshetri, N. (2021). Blockchain and sustainable supply chain management in developing countries. International Journal of Information Management, 60. doi:10.1016/j.ijinfomgt.2021.102376.

Li, C.Z., Chen, Z., Xue, F., Kong, X.T.R., Xiao, B., Lai, X. and Zhao, Y. (2021). A blockchain- and IoT-based smart product-service system for the sustainability of prefabricated housing construction. Journal of Cleaner Production, 286. doi:10.1016/j.jclepro.2020.125391.

Lund, E.H., Jaccheri, L., Li, J., Cico, O. and Bai, X. (2019). Blockchain and sustainability: A systematic mapping study. In 2019 IEEE/ACM 2nd International Workshop on Emerging Trends in Software Engineering for Blockchain (WETSEB) (pp. 16–23). IEEE.

Mahyuni, L.P., Adrian, R., Darma, G.S., Krisnawijaya, N.N.K., Dewi, I.G.A.A.P. and Permana, G.P.L. (2020). Mapping the potentials of blockchain in improving supply chain performance. Cogent Business and Management, 7(1). doi:10.1080/23311975.2020.1788329.

Mangla, S.K., Kazancoglu, Y., Ekinci, E., Liu, M., Özbiltekin, M. and Sezer, M.D. (2021). Using system dynamics to analyze the societal impacts of blockchain technology in milk supply chainsrefer. Transportation Research Part E: Logistics and Transportation Review, 149. doi:10.1016/j.tre.2021.102289.

Marsal-Llacuna, M. (2020). The people's smart city dashboard (PSCD): Delivering on community-led governance with blockchain. Technological Forecasting and Social Change, 158. doi:10.1016/j.techfore.2020.120150.

Massaro, M., Dal Mas, F., Chiappetta Jabbour, C.J. and Bagnoli, C. (2020). Crypto-economy and new sustainable business models: Reflections and projections using a case study analysis. Corporate Social Responsibility and Environmental Management, 27(5): 2150–2160. doi:10.1002/csr.1954.

Min, S., Zacharia, Z.G. and Smith, C.D. (2019). Defining supply chain management: In the past, present, and future. Journal of Business Logistics, 40(1): 44–55. doi:10.1111/jbl.12201.

Mishra, L. and Kaushik, V. (2021). Application of blockchain in dealing with sustainability issues and challenges of financial sector. Journal of Sustainable Finance and Investment. doi:10.1080/20430795.2021.1940805.

Mohsin, M., Naseem, S., Zia-ur-Rehman, M., Baig, S.A. and Salamat, S. (2020). The crypto-trade volume, GDP, energy use, and environmental degradation sustainability: An analysis of the top 20 crypto-trader countries. International Journal of Finance and Economics. doi:10.1002/ijfe.2442.

Moore, J.E., Mascarenhas, A., Bain, J. and Straus, S.E. (2017). Developing a comprehensive definition of sustainability. Implementation Science, 12(1): 1–8. Doi:10.1186/s13012-017-0637-1.

Niu, B., Shen, Z. and Xie, F. (2021). The value of blockchain and agricultural supply chain parties' participation confronting random bacteria pollution. Journal of Cleaner Production, 319. doi:10.1016/j.jclepro.2021.128579.

Okoli, C. (2015). A guide to conducting a standalone systematic literature review. Communications of the Association for Information Systems, 37(1): 879–910.

Parmentola, A., Petrillo, A., Tutore, I. and De Felice, F. (2021). Is blockchain able to enhance environmental sustainability? A systematic review and research agenda from the perspective of sustainable development goals (SDGs). Business Strategy and the Environment. doi:10.1002/bse.2882.

Poberezhna, A. (2018). Addressing water sustainability with blockchain technology and green finance. Transforming Climate Finance and Green Investment with Blockchains (pp. 189–196). doi:10.1016/B978-0-12-814447-3.00014-8.

Quayson, M., Bai, C. and Sarkis, J. (2021). Technology for social good foundations: A perspective from the smallholder farmer in sustainable supply chains. IEEE Transactions on Engineering Management, 68(3): 894–898. doi:10.1109/TEM.2020.2996003.

Rainero, C. and Modarelli, G. (2021). Food tracking and blockchain-induced knowledge: A corporate social responsibility tool for sustainable decision-making. British Food Journal. doi:10.1108/BFJ-10-2020-0921.

Rana, R.L., Tricase, C. and De Cesare, L. (2021). Blockchain technology for a sustainable agri-food supply chain. British Food Journal, doi:10.1108/BFJ-09-2020-0832.

Rejeb, A. and Rejeb, K. (2020). Blockchain and supply chain sustainability. [Blockchain i zrównoważoność łańcucha dostaw] Logforum, 16(3): 363–372. doi:10.17270/J.LOG.2020.467.

Robertson, M. (2017). Sustainability Principles and Practice (2nd ed.). Routledge.

Saberi, S., Kouhizadeh, M., Sarkis, J. and Shen, L. (2019). Blockchain technology and its relationships to sustainable supply chain management. International Journal of Production Research, 57(7): 2117–2135. doi:10.1080/00207543.2018.1533261.

Schinckus, C. (2020). The good, the bad and the ugly: An overview of the sustainability of blockchain technology. Energy Research and Social Science, 69: 101614.

Schulz, K.A., Gstrein, O.J. and Zwitter, A.J. (2020). Exploring the governance and implementation of sustainable development initiatives through blockchain technology. Futures, 122. doi:10.1016/j.futures.2020.102611.

Scoones, I. (2016). The politics of sustainability and development. Annual Review of Environment and Resources, 41: 293–319.

Shojaei, A., Ketabi, R., Razkenari, M., Hakim, H. and Wang, J. (2021). Enabling a circular economy in the built environment sector through blockchain technology. Journal of Cleaner Production, 294. doi:10.1016/j.jclepro.2021.126352.

Song, L., Wang, X. and Merveille, N. (2020). Research on blockchain for sustainable E-agriculture. Paper presented at the 2020 IEEE Technology and Engineering Management Conference, TEMSCON 2020, doi:10.1109/TEMSCON47658.2020.9140121.

Soni, G., Mangla, S.K., Singh, P., Dey, B.L. and Dora, M. (2021). Technological interventions in social business: Mapping current research and establishing future research agenda. Technological Forecasting and Social Change, 169. doi:10.1016/j.techfore.2021.120818.

Svetec, E., Nad, L., Pasicko, R. and Pavlin, B. (2019). Blockchain application in renewable energy microgrids: An overview of existing technology towards creating climate-resilient and energy independent communities. Paper presented at the International Conference on the European Energy Market, EEM, 2019-September doi:10.1109/EEM.2019.8916292.

Tanev, S. and Sandstrom, G. (2020). Innovating in times of crisis. Technology Innovation Management Review, 10(6): 3–4. doi:10.22215/timreview/1363.

Tham, A. and Sigala, M. (2020). Roadblock(chain): Bit(coin)s for tourism sustainable development goals? Journal of Hospitality and Tourism Technology, 11(2): 203–222. doi:10.1108/JHTT-05-2019-0069.

Tian, Z., Zhong, R.Y., Vatankhah Barenji, A., Wang, Y.T., Li, Z. and Rong, Y. (2021). A blockchain-based evaluation approach for customer delivery satisfaction in sustainable urban logistics. International Journal of Production Research, 59(7): 2229–2249. doi:10.1080/00207543.2020.1809733.

Tiscini, R., Testarmata, S., Ciaburri, M. and Ferrari, E. (2020). The blockchain as a sustainable business model innovation. Management Decision, 58(8): 1621–1642. doi:10.1108/MD-09-2019-1281.

Tiwari, P., Ilavarasan, P.V. and Punia, S. (2019). Content analysis of literature on big data in smart cities. Benchmarking, 28(5): 1837–1857. doi:10.1108/BIJ-12-2018-0442.

Tozanlı, Ö., Kongar, E. and Gupta, S.M. (2020). Trade-in-to-upgrade as a marketing strategy in disassembly-to-order systems at the edge of blockchain technology. International Journal of Production Research, 58(23): 7183–7200. doi:10.1080/00207543.2020.1712489.

Tseng, M. -., Bui, T. -., Lim, M.K., Tsai, F.M. and Tan, R.R. (2021). Comparing world regional sustainable supply chain finance using big data analytics: A bibliometric analysis. Industrial Management and Data Systems, 121(3): 657–700. doi:10.1108/IMDS-09-2020-0521.

Upadhyay, A., Mukhuty, S., Kumar, V. and Kazancoglu, Y. (2021). Blockchain technology and the circular economy: Implications for sustainability and social responsibility. Journal of Cleaner Production, 293. doi:10.1016/j.jclepro.2021.126130.

Wong, L., Leong, L., Hew, J., Tan, G.W. and Ooi, K. (2020). Time to seize the digital evolution: Adoption of blockchain in operations and supply chain management among Malaysian SMEs. International Journal of Information Management, 52. doi:10.1016/j.ijinfomgt.2019.08.005.

Yousif Alsharidah, Y.M. and Alazzawi, A. (2020). Artificial intelligence and digital transformation in supply chain management A case study in Saudi companies. Paper presented at the 2020 International Conference on Data Analytics for Business and Industry: Way Towards a Sustainable Economy, ICDABI 2020. doi:10.1109/ICDABI51230.2020.9325616.

CHAPTER 6

Blockchain x Green Open Innovation

Evidence and Trends for Sustainability

Suelen Marostica, Ronise Suzuki, Andreia de Bem Machado, Gertrudes Aparecida Dandolini* and *João Artur de Souza*

1. Introduction

With the current climate emergency and concerns regarding the impact of economic activities on the environment, green innovation is particularly urgent and necessary. It is a motivating aspect of innovation that leads to a competitive advantage for companies that wish to address contemporary economic, environmental, and social issues (D'Agostino, 2020). Innovative solutions such as sustainable technologies, policies, and new business models are needed to address complex challenges and uncertainties, (Howson, 2021).

However, it should be noted that the process of green innovation is arduous and faces significant information asymmetry, mainly due to the maturity level of green technologies (Yuan et al., 2021). Due to the

Department of Engineering and Knowledge Management, Federal University of Santa Catarina.
Emails: marostica.suelen@gmail.com; ronisesuzuki@gmail.com; Gertrudes.dandolini@ufsc.br; joao.artur@ufsc.br
* Corresponding author: andreiadebem@gmail.com

complexity of knowledge required, the crucial role of stakeholders, and the technologies that can be adopted, green innovation presents additional challenges to the usual difficulties of the innovation process (D'Agostino, 2020).

Open innovation is one approach that can help overcome such challenges, and companies should work together to ensure that sustainable development is possible (Mubarak et al., 2021) without being burdened with the cost-benefit effects of doing green innovation. In this context, companies are increasingly relying on the open innovation approach so that their internal and external knowledge frontiers are extended through collaboration, cooperation, and sharing of knowledge and experiences with competitors and external partners (Chesbrough, 2003, 2006; De Marchi, 2012; De Marchi and Grandinetti, 2013).

With the perspective that this environment, where the green innovation process is essential and the facilitation of this process by open innovation, digital technologies are being used for several applications relevant to economic exchange and environmental sustainability.

In this sense, Dahlander et al. (2021) discuss that, in general, technological developments in the last decade have changed the "when" and "how" organizations can rely on innovation, as advances have occurred in essential technologies such as artificial intelligence, machine learning, big data, cloud solutions, advanced robotics, and blockchain. According to Marshall (2021), these digital technologies have expanded practices and ideas around sharing knowledge and ideas that favor open and collaborative innovation processes. In other words, these digital technologies are transforming a wide variety of innovation processes.

Technologies such as blockchain can be helpful by providing agility and trust between human agents in open innovation processes (Pazaits, 2020; Patrickson, 2021). Both open innovation and blockchain are concepts that shape the way organizations work together and share information.

Open innovation has been extensively researched over the past ten years, while blockchain has attracted the attention of innovation researchers in recent years, as has application in green open innovation processes.

This study aims to investigate the application of blockchain to open innovation and green innovation processes in literature, seeking to identify trends for future research through a rigorous and replicable systematic review process (Massaro et al.,2016). Thus, this chapter contributes to reflecting on the following research questions:

1. Is it possible to identify evidence for stimulating green innovation in blockchain applications for open innovation?
2. Which applications of blockchain are appropriate for green innovation?
3. What should be the direction of future research in this domain?

This chapter discusses the concepts of open innovation, green innovation, and blockchain to identify the evidence and trends of applying blockchain for open innovation and green innovation processes, highlighting the challenges and benefits in the process.

2. The Open Innovation Approach as a Pathway to Green Innovation

The concept of open innovation means "a paradigmatic shift in the way firms approach innovation activities" (Pazaitis,2020, p. 3). Also, according to this author, in open innovation, knowledge flows and market trajectories of internal and external origin merge into innovative business activities, triggering more significant implications for the respective business practices and structure. Firm boundaries become less defined, and network-based forms of organization gain prominence in successful strategies.

Chesbrough and Bogers (2014) define *open innovation* as a distributed innovation process involving intentional knowledge flows across organizational boundaries for monetary or non-monetary reasons. There are two main directions in the flow of knowledge: outside-in and inside-out. Open innovation means sharing and distributing knowledge, fostering cooperation, and enhancing distributed innovation processes.

The open innovation strategy adopts the premise of open system theory, emphasizes a company's interaction with external stakeholders to improve its innovation performance, and enables the development of an innovation ecosystem, with networks of external partners for new knowledge development (Lichtenthaler and Lichtenthaler, 2009; Salampasis and Mention, 2019; Murabak and Petraite, 2020).

On the other hand, open innovation is arguably linked to more radical transformations in how production processes incorporate knowledge to create more socially meaningful outcomes and stimulate green innovations in open innovation research (Latusek-Jurczak and Prystupa-Rzadca, 2014; Koh et al., 2020; Mubarak et al., 2021).

In this context, trust among inter-organizational partners and teams is a success factor, as collaborations and partnerships are undertaken -predominantly- to enable intentional knowledge input and output to innovate (Chesbrough and Bogers, 2014).

The essence of open innovation lies in collaboration with external stakeholders to obtain external knowledge, which plays a key role in developing mutual competence. This competence helps a firm improve its innovation activities' efficiency and effectiveness (Gassmann and Enkel, 2004; Chesbrough and Crowther, 2006; Lichtenthaler, 2011; Podmetina et al., 2018; Murabak and Petraite, 2020).

According to Bogers, Chesbrough, and Strand (2020), the concept of open innovation is very useful for effectively describing and coordinating activities related to sustainability goals and stimulating green innovation processes. According to the authors, many companies in various industries will face challenges, often linked to the imperative of sustainability in the future.

The open innovation approach enables intelligent experimentation to unlock pathways to green innovation by mitigating risks, costs, and sharing knowledge, experience, and learning (Chesbrough, 2010; Blank, 2013; Borges et al., 2020).

For Murabak et al. (2021), realizing and executing green innovation requires a significant restructuring and leap over existing technologies and processes. Organizations improve their knowledge base to reshape all stakeholders' interaction, priorities, and behavior to sustainable and green innovation measures. Furthermore, digital technologies like blockchain can improve the performance of open green innovation, but need to be applied at scale.

3. Blockchain: Concepts and Possibilities for Green Open Innovation

Blockchain consists of a distributed ledger technology. The technology enables the recording and sharing of information in blocks and connects them in chronological sequence in a specific, encrypted data structure through computational effort (Nakamoto, 2008). The transaction process is transparent, traceable, and tamper-proof (Yang and Li, 2020).

The security of a blockchain is, derived from collective inspection due to the need to obtain consensus on multiple permanent records in a peer-to-peer system. This way, it avoids the need for additional authentication from trusted third parties (Patrickson, 2021).

According to Kassem (2021), storing information in a strictly sequential and decentralized manner makes the blockchain-based recording system reliable in recording and storing information. As part of the natural process of entropy, information can dissipate over time in virtually any ecosystem that stores it. As such, blockchain has features and advantages to solve the problems of sharing information resources, as shown in Box 1.

The adoption of blockchain technology changes the way stakeholders collaborate within an organization and between organizations. Organizations can share their knowledge with other companies more securely, with more control and benefit in the process. In turn, companies can focus on their core business and use knowledge from outside based on their needs (Li et al., 2018).

Box 1. Features and advantages of blockchain.

Features	**Benefit**
Decentralization	All network participants (nodes) are involved in data maintenance. There is no control of the data flow through a third party.
Immutability	Records can hardly be tampered with by malicious parties or modified by unauthorized parties.
Reliability	Reliability allows people to communicate and cooperate without worrying about integrity issues through the consensus mechanism.
Transparency and traceability	All data is permanently stored in the blocks and cannot be deleted. In addition, all data audit trails are transparent and supervised by all network participants.

Source: Elaborated from Nakamoto (2008), Yang e Li (2020) e Patrickson (2021).

In open innovation processes, smart contracts can be an efficient and secure means for knowledge sharing. According to Yang and Li (2020), smart contracts automatically execute codes with flexible and programmable features and do not depend on any intermediary. They are created and triggered by how entities send transactions, avoiding unilateral violation of rules. In smart contracts, the entity's transaction process is made public, and the underlying consensus mechanism of the blockchain ensures the accuracy and consistency of transaction results. In addition, smart contracts make it possible to monitor and manage knowledge usage and apply penalties accordingly (Nyame et al., 2020).

Murabak and Petraite (2020) discuss that the power of distributed ledgers and smart contracts promise to eliminate friction and radically reduce transaction costs while increasing security, verifiability, and transparency. Because blockchain is an available technology, it enables greater agility and distribution of power through decentralization (Narayan and Tidström, 2019), which can create favorable conditions for open innovation.

Transitioning companies can use blockchain-powered open innovation to test new ideas and allow technology to shape the development process (Seulliet, 2016; Treiblmaier and Sillaber, 2020). Virtually anything of value can be committed and traded on a blockchain network (Iansiti and Lakhani, 2017). By this logic, blockchain facilitates the flow of communication, provides security for authorship recognition, and strengthens trust that is essential for sharing and creating new knowledge, acting as an agent to support better knowledge sharing in organizations and between organizations (Nyame et al., 2020) aimed at open innovation.

New digital technologies, such as blockchain, are inherently dynamic and malleable and enable knowledge sharing in new ways to produce new products and services and allow organizations to experience rapid and radical change more frequently than previously considered (Yoo 2012; Li et al., 2018; Dahlander et al., 2021; Patrickson, 2021). Moreover, when combined with other technologies such as sensors (IOTs and Cyber-Physical Systems), blockchain can provide a sophisticated platform for recording and sharing data and controlling the entire production chain, which enables better efficiency in the use of natural resources and more assertive decision making, positively impacting business performance (Manesh et al., 2020; Mubarak et al., 2021).

4. Methodology

The present study conducts a Systematic Literature Review (SLR) to synthesize blockchain's contribution to open green innovation. The study follows the SLR guidelines (Tranfield et al., 2003) and the Preferred Reporting Items for Systematic Reviews and Meta-Analyses (PRISMA) protocols (PRISMA, 2020) to ensure the robustness of the SLR and the reliability of the results.

The research is exploratory with a qualitative approach and has two stages. The first stage is focused on the survey and analysis of the body of technical-scientific knowledge produced on blockchain x, and available innovation-seeking evidence of applications that stimulate green innovation or more sustainable processes. The second is focused specifically on the survey and analysis of the body of technical-scientific knowledge produced on blockchain x green innovation-seeking evidence of blockchain applications for this purpose.

The search was conducted in the databases SCOPUS, Science Direct, and Web of Science in December 2021, focusing on blockchain x open innovation (step 1) and blockchain x green innovation (step 2).

Given that the aim of the research was the applications of blockchain for open innovation and green innovation, the four exclusion criteria were applied. We ensured that articles (1) were not duplicated from the different databases, (2) had adherence with the research objective (it was verified the application of blockchain for open innovation processes and green innovation), (3) were discussing the relationship between the search terms and not just citing them and, (4) were restricted to applied research. It is worth mentioning that there was no delimitation by year of publication in the searches.

Thus, a search corpus of 21 articles published between 2018 and 2021 for blockchain x open innovation and a corpus of six articles published between 2020 and 2021 for blockchain x green innovation were defined.

The systematic review helps identify an underlying trend in keywords and possible avenues for the evolution of studies in the relevant areas (Box 2).

Box 2. Results for each search stage of the search.

Stage of the Search	Search String	Initial Result	The Result after the Exclusion Procedure	Bases Searched
First stage	"open innovation" AND blockchain	118 articles	21	Scopus Science Direct Web of Science
Second stage	"green innovation" AND blockchain	57 articles	6	Scopus Science Direct Web of Science

Source: Elaborated by the authors.

With the software VOSviewer version 1.6.17, it was possible to generate a graph that illustrates the main trends in the research field. The keywords represent the unit of analysis; they are terms cited by the authors at least once in each document. Synonymous terms were excluded. After the articles were read thoroughly, they were sorted according to the application of blockchain to stimulate open innovation or green innovation, respectively under the heads, "Open Innovation and Blockchain: what do scientific articles say?" and "Blockchain and Green Innovation - possible paths."

5. Open Innovation and Blockchain - What do Scientific Articles Say?

Keyword analysis highlighted research trends on the topic, as seen in Figure 1. It is noteworthy that research on blockchain applications for open innovation began to grow in 2019. However, in August 2018, a publication was made on the topic. The highest concentration of published articles on the topic is between 2020 and 2021.

As discussed in Murabak et al. (2020) and Marshall (2021), digital technologies such as the internet of things, blockchain, and cyber-physical systems represent a paradigm shift in the shape of business processes and innovations.

As observed in Figure 1, the application of blockchain for open innovation is an expanding field of research with numerous application possibilities. Marshal (2021) discussed that the development of technologies using blockchain is on the rise. It is a distributed accounting technology that employs chain transactions with continuously growing blocks and protects all transactions or data using encryption.

Blockchain technology enables co-ownership, reliability, and efficiency for open innovation purposes. While the information flowing

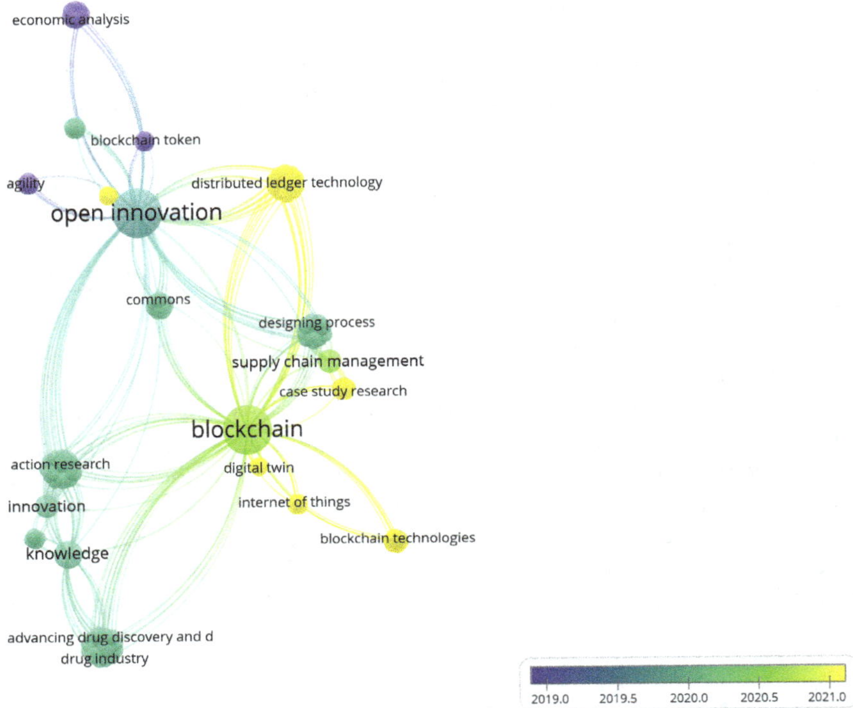

Figure 1. Time map of the occurrence of keywords for the combined terms "open innovation" and "blockchain".
Source: Research data.

between participants has previously been managed on the servers of service providers or other agents, such as governments, the emergence of blockchain technology allows for the storage and distribution of information between participants, establishing direct connections between them, resulting in a reduction of the business value chain (Park et al., 2020).

As illustrated in Figure 2, thirteen research clusters were found around blockchain and open innovation. It was observed that the cluster with the highest adherence to the search terms was strongly related to the terms of antifragility systems, token systems, and supply chains. This trend is evident in the studies pointing to the use of blockchain applied to supply chain management by Rahmanzadeh et al. (2020), Li et al. (2021), Passareli et al. (2021), and Teodorescu and Korchagina (2021).

Another point of note is that by the map of terms, it is observed that in 2019 the studies on the application of blockchain for open innovation were more related to agility issues and the economic effects of these practices. From 2020 the studies are directed to new blockchain technologies,

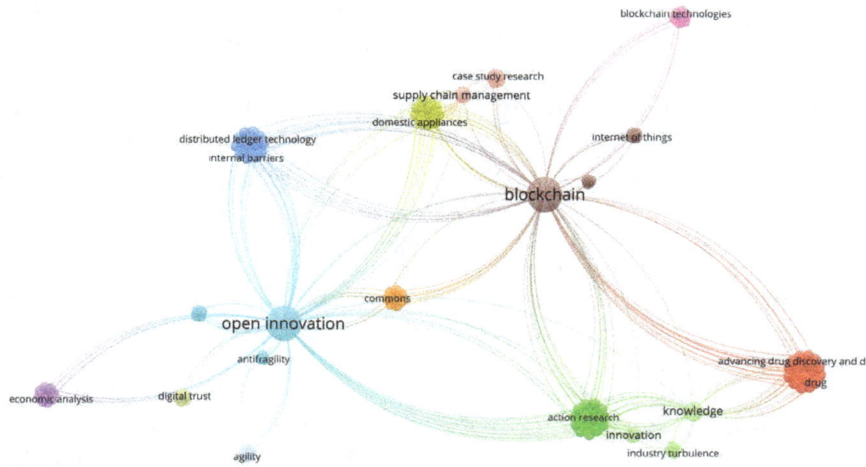

Figure 2. Clusters identified with keywords for the combined terms "open innovation" and "blockchain.

Source: Research data.

supply chains, and more robust security systems to mitigate weaknesses and ensure digital trust in sharing data and information. An interesting factor is the trust aspect. In the area of open innovation, trust is a necessary condition for collaboration, contributing to the free exchange of information, and avoiding the fear of uncontrolled information disclosure (Park et al., 2020; Mubarak et al., 2021; Li, 2021; Passereli et al., 2021; Teodorescu and Korchagina, 2021).

6. Key Approaches to Applying Blockchain in Open Innovation Processes

Besides being an innovative technology that can dramatically improve the efficiency, transparency, and verifiability of supply chains (Rahmanzadeh et al., 2020; Li et al., 2021; Passarelli et al., 2021 and Teodorescu and Korchagina, 2021), blockchain technology enables disruptive digital innovations (Christensen, 2013). Pazaits (2020) and Patrickson (2021) also claim it brings agility in open innovation processes.

On the other hand, intellectual property issues and protection of innovators' rights are significant concerns that sometimes prevent companies from engaging in open innovation initiatives. Therefore, mechanisms are needed that allow opportunities to be discovered and exploited collaboratively, reducing uncertainty (Rahmanzadeh et al., 2020), which can be facilitated by blockchain technology.

Park (2021) highlights the widespread adoption of open innovation within the blockchain industry. The author emphasizes how these innovation practices have significantly augmented the knowledge base in blockchain technology, fostering its growth and fostering new developments. Pazaits (2020) and Patrickson (2021) support this by suggesting that these activities also enhance agility within open innovation processes.

Of the 21 articles analyzed in-depth, 19 deal with the application of blockchain for open innovation in companies, only one focuses on a government application, and only one deals with the application for open innovation processes in university and company interaction, as can be seen in Box 3.

In Box 4, we observe that the main discussions presented in these articles point to digital trust, digital collaboration, and efficiency in the management process of open innovation. Other points highlighted in the publications are the importance of applying blockchain in supply chain management for sharing information and innovations in the sector. In addition to these aspects, the collaborative approach to developing new systems, efficiency in payments, and support in changing organizational culture were observed.

According to Benitez et al. (2021), there is a global trend towards open innovation ecosystems that allow data owners to have better control over their data and privacy, choosing whether/what and with whom to share/trade-specific data streams. Therefore, using blockchain for the open innovation process will enable a secure way to register ideas from the beginning and introduce a decentralized incentive system that can encourage innovators to develop their ideas further and be appreciated, recognized, and remunerated for their work.

Digital trust is a problem for open innovation, and digital technologies such as blockchain can improve these requirements (Chesbrough, 2020; Rajput and Singh, 2019; Murabak et al., 2020).

It is worth noting that among the articles analyzed in-depth, five articles presented the application of blockchain for sustainability processes; two articles focused on analyzing the efficiency and optimization of resource use in supply chains (Mubarak et al. (2021); Passarelli et al. (2021) one article focused on smart government (Benítez-Martínez, 2021), one article focused on sustainable buildings (Li et al., 2021), and one article focused on how digital technologies can support sustainable activities (Pazaitis, 2020).

Box 3. Summary of the articles analyzed.

Title	Research Objective	Application	Authors
Blockchain tokens and the potential democratization of entrepreneurship and innovation.	Explore how blockchain reshapes fundraising, investment, community building, and open source.	ENTERPRISE - Use blockchain for digital trust and digital collaboration for open innovation with information sharing and funding attraction.	Chen (2018)
An innovative technology: Augmented reality-based information systems.	Analyze next-generation artificial reality systems related to aerospace and defense, industrial, education, medical, and gaming sectors.	ENTERPRISE - Use blockchain to manage digital collaboration and trust for systems development.	Aslan et al. (2019)
Digitalization needs a cultural change–examples of applying Agility and Open Innovation to drive the digital transformation.	Develop a strategy for cultural change through blockchain and open innovation.	ENTERPRISE - Using blockchain to manage knowledge sharing and digital collaboration for open innovation.	Burchardt and Maisch (2019)
Open source adoption strategy.	Characterize under what conditions organizations adopt open source software and examine whether these adoption patterns are socially beneficial.	ENTERPRISE - Use of blockchain to manage digital collaboration and digital trust for open source systems development.	Katsamakas and Xin (2019)
Sharing my way to success: A case study on developing entrepreneurial ventures using social capital in an OSS community.	Compare the behaviors of entrepreneurs and non-entrepreneurs to investigate how entrepreneurs build social capital within an open community.	ENTERPRISE - Entrepreneurial behavior uses blockchain to manage knowledge sharing and digital collaboration for open innovation and digital collaboration.	Larsson et al. (2019)
Exploring the influence of common game elements on ideation output and motivation.	Investigate the incentive effect of gamification elements in producing online idea contests.	ENTERPRISE - Using blockchain for gamification and to manage open innovation.	Zimmerling et al. (2019)

Box 3 contd. ...

...Box 3 contd.

Title	Research Objective	Application	Authors
How the effect of opportunity discovery on innovation outcome differs between DIY laboratories and public research institutes: The role of industry turbulence and knowledge generation in the case of Singapore.	Examine how opportunity discovery directly affects innovation outcomes and how industry turbulence moderates the effect of opportunity discovery on innovation outcomes of DIY Maker Labs relative to Public Research Institutes.	UNIVERSITY - COMPANY INTERACTION Use of blockchain to manage knowledge sharing and digital collaboration for open innovation.	Cheah et al. (2020)
Integrated innovative product design and supply chain tactical planning within a *blockchain* platform.	Optimize tactical decisions according to supply chain objectives and open innovation considerations.	ENTERPRISE - Use blockchain to manage digital collaboration to increase efficiency in logistics and ensure digital trust.	Rahmanzadeh et al. (2020)
Breaking the chains of open innovation: post-Blockchain and the case of Sensorica.	Explore the internal dynamics of commons-based peer production (CBPP) to enable and support innovative, sustainable, and responsible activities.	ENTERPRISE - Using blockchain to manage knowledge sharing, efficiency, and digital collaboration for open innovation.	Pazaitis (2020)
The effects of the early mover (dis) advantages and knowledge spillover on blockchain startups' funding and innovation performance.	Examine how the timing of a startup's entry into the blockchain industry affects its attraction of venture capital funding and innovation performance. In addition, this study examines whether the startup's knowledge spillover activities moderate the relationship between the timing of entry and subsequent performance.	ENTERPRISE - Use blockchain for digital trust and digital collaboration for open innovation with information sharing and funding attraction.	Park et al. (2020)

Box 3 contd. ...

...Box 3 contd.

Title	Research Objective	Application	Authors
Enhanced Lightning Network (off-chain)-based micropayment in IoT ecosystems.	Characterize the requirements that need to be met to support (micro) payment in IoT adequately and to what extent different blockchain technologies can meet these requirements.	ENTERPRISE - Using blockchain for efficiency in means of payments and reducing transaction costs and for digital trust.	Robert et al. (2020)
Industry 4.0 technologies, digital trust, and technological orientation: What matters in open innovation?	Investigate the role of technological orientation and technological absorptive capacity in the association between digital trust and open innovation.	ENTERPRISE - Use blockchain to ensure digital trust and innovation management, focusing on data security.	Mubarak and Petraite (2020)
Critical factors in the startup phase of collaborative foresight.	Investigate the incentive effect of gamification elements in producing online idea contests.	ENTERPRISE - Use of blockchain for gamification and to manage open innovation.	Gattringer and Wiener (2020)
A blockchain-and IoT-based smart product-service system for the sustainability of prefabricated housing construction.	Development of an intelligent platform for resource management in prefabricated buildings.	ENTERPRISE - Use blockchain to manage digital collaboration, increase logistics collaboration efficiency, and ensure digital trust.	Li et al. (2021)

Box 3 contd. ...

...Box 3 contd.

Title	Research Objective	Application	Authors
A neural blockchain for a tokenizable e-Participation model.	Develop an e-participation model that uses a tokenizable system of the actions and processes performed by citizens in participatory processes providing incentives to promote greater participation in public affairs. In order to develop a sustainable, scalable, and resilient e-participation system, a new blockchain concept, which organizes blocks like a neural system, is combined with the implementation of a virtual token to reward participants.	GOVERNMENT - Using blockchain for digital participation and collaboration for open innovation.	Benítez-Martínez et al. (2021)
Distributed ledger technology as a catalyst for open innovation adoption among small and medium-sized enterprises.	Investigate the barriers to open innovation currently faced by small and medium-sized enterprises (SMEs) that DLT can address.	ENTERPRISE - Using blockchain to manage knowledge sharing and digital collaboration for open innovation.	Hashimy et al. (2021)
A blockchain-based digital twin sharing platform for reconfigurable socialized manufacturing resource integration.	Develop a blockchain-based digital twin sharing platform to enable software copyright protection and simplify the integration of heterogeneous manufacturing resources in decentralized and distributed environments.	ENTERPRISE - Using blockchain to secure digital trust, share information, to preserve and recover digital twin copyrights in a distributed and decentralized SMC network.	Li et al. (2021)

Box 3 contd. ...

...Box 3 contd.

Title	Research Objective	Application	Authors
How Industry 4.0 technologies and open innovation can improve green innovation performance?	Investigate the impact of blockchain technology on green supply chain practices to promote pro-environmental configurations in manufacturing company supply chains.	ENTERPRISE - Using blockchain to increase efficiency in green supply chains.	Mubarak et al. (2021).
LiSC Model: an innovative paradigm for Liquid Supply Chain.	Propose and describe a new supply chain construct called: Liquid Supply Chain (LiSC).	ENTERPRISE- Use blockchain to manage digital collaboration to increase efficiency in logistics and ensure digital trust.	Passarelli et al. (2021)
What do blockchain technologies imply for digital creative industries?	Study the capability and potential of blockchain technologies and the social and industrial contexts in which these technologies can be applied.	ENTERPRISE - Blockchain to manage knowledge sharing, efficiency, and digital collaboration for open innovation.	Patrickson (2021)
Applying Blockchain in the Modern Supply Chain Management: Its Implication on Open Innovation.	Analyze the practice of blockchain implementation in logistics and supply chain management.	ENTERPRISE - Use blockchain to manage digital collaboration to increase efficiency in logistics and ensure digital trust.	Teodorescu and Korchagina (2021)

Source: Elaborated by the authors.

7. Blockchain and Green Innovation-possible Paths

Blockchain applications for green innovation processes is a field of research that is still expanding. Considering the numerous potential applications of blockchain, a search was conducted with the specific goal of finding scientific research on the application of blockchain to stimulate green innovation. Figure 3 evidences that the publications on blockchain application and green innovation are recent and started in 2020.

As illustrated in Figure 3, as of 2021, a focus on blockchain applications in green innovation, such as clean energy, smart agriculture and smart city, green building, as well as on exploring more robust algorithms for

Box 4. Main discussions identified in the analyzed articles.

Topics Covered	Authors
Digital trust	Chen (2018); Aslan et al. (2019); Burchardt and Maisch (2019); Katsamakas and Xin (2019); Larsson et al. (2019); Rahmanzadeh et al. (2020); Pazaitis (2020); Robert et al. (2020); Li et al. (2021); Passarelli et al. (2021); Teodorescu and Korchagina (2021)
Digital collaboration	Chen (2018); Aslan et al. (2019); Burchardt and Maisch (2019); Katsamakas and Xin (2019); Larsson et al. (2019); Cheah (2020); Gattringer and Wiener (2020); Park et al. (2020); Pazaitis (2020); Rahmanzadeh et al. (2020); Robert et al. (2020); Benítez-Martínez et al. (2021); Hashimy et al. (2021); Passarelli et al. (2021); Patrikson (2021) Teodorescu and Korchagina (2021)
Efficiency in the process of open innovation management	Chen (2018); Aslan et al. (2019); Burchardt and Maisch (2019); Katsamakas and Xin (2019); Larsson et al. (2019); Zimmerling et al. (2019); Cheah (2020); Gattringer and Wiener (2020); Park et al. (2020); Pazaitis (2020); Rahmanzadeh et al. (2020); Robert et al. (2020); Benítez-Martínez et al. (2021); Hashimy et al. (2021); Passarelli et al. (2021); Patrickson (2021) Teodorescu and Korchagina (2021)
System development	Aslan et al. (2019); Katsamakas and Xin (2019); Li et al. (2021)
Cultural change	Burchardt and Maisch, (2019); Benítez-Martínez (2021)
Supply chain efficiency	Rahmanzadeh et al. (2020); Li et al. (2021); Mubarik et al. (2021); Passarelli et al. (2021); Teodorescu and Korchagina (2021)
Means of payments	Park et al. (2020); Robert et al. (2020)

Source: Elaborated by the authors.

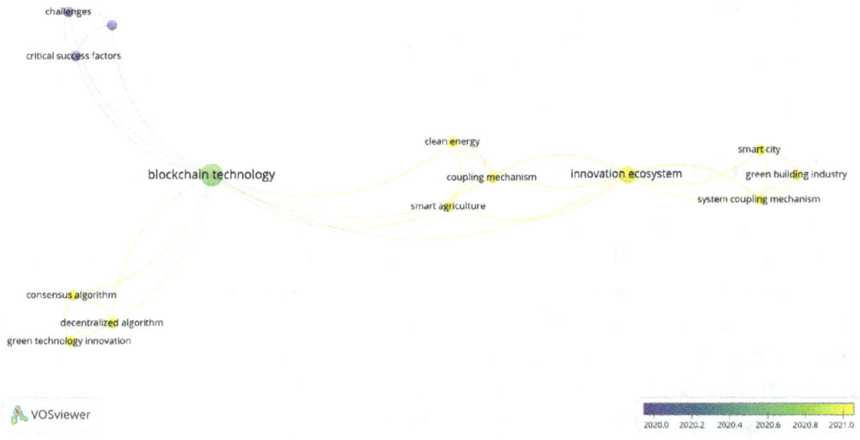

Figure 3. Time map of keyword occurrence for the combined terms "green innovation" and "blockchain.

Source: Research data.

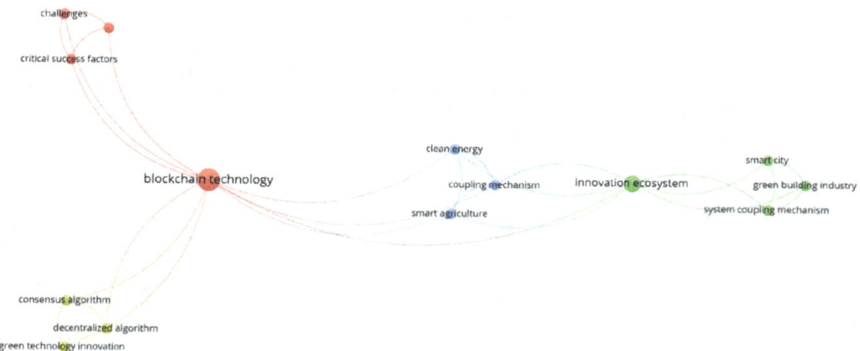

Figure 4. Clusters identified with keywords for the combined terms "green innovation" and "blockchain".
Source: Research data.

digital trust, decentralization, and green technological innovation can already be identified.

Four research clusters were identified around the terms blockchain and green innovation, as illustrated in Figure 4. It was observed that the cluster with the highest adherence to the search terms was strongly related to the terms of trustworthiness, decentralization, change, and critical success factors. These issues are addressed to a greater or lesser degree in all the articles analyzed, given that all over the world, companies are collaborating with reliable partners to enrich their knowledge and research base to improve their green products and processes (Li et al., 2021; Mubarak et al. 2021).

The main discussions presented in these papers point to issues of digital trust, digital collaboration, costs, uncertainties, the volume of information, and complexity and maturity of green innovations (Hou et al., 2021; Li, 2021; Mubarak et al., 2021). Another point of emphasis is how blockchain can mitigate these problems and improve the performance of green innovations (Hoswon, 2021; Mubarak et al., 2021).

Green innovation is a type of economic activity with high uncertainty and risk. This requires that companies have a specific potential to continuously obtain, rebuild, and integrate internal and external resources and the use of digital technologies to secure and enhance green innovation performance (Jiang and Zheng, 2021).

The systematic review, summarized in Box 5, shows that the main applications of blockchain for green innovation are related to (a) new technologies to stimulate green innovation processes in a collaborative and shared way and (b) search for new technologies to mitigate problems of natural resources use (Hou et al., 2021; Li et al., 2021; Jiang and Zheng, 2021; Mubarak et al., 2021).

Box 5. Summary of analyzed articles.

Title	Research Objective	Application	Author
The key challenges and critical success factors of blockchain implementation: Policy implications for Singapore's maritime industry.	Identify the key challenges and critical success factors (CSFs) of blockchain implementation in the maritime industry.	Maritime industry	Zhou et al. (2020)
Coupling mechanism and development prospect of the innovative clean energy ecosystem in smart agriculture based on blockchain.	Analyze the coupling mechanism and development perspective of the clean energy innovation ecosystem under blockchain-based smart agriculture.	Smart agriculture and renewable energy generation	Hou et al. (2021)
Distributed degrowth technology: Challenges for blockchain beyond the green economy.	Analyze the challenges of using blockchain technology to facilitate the transition from degrowth to sustainability in three essential areas: (1) equitable resource (re) distribution, (2) ecological regeneration, and (3) decolonization.	Degrowth to sustainability	Howson (2021)
The coupling mechanism of green building industry innovation ecosystem based on blockchain smart city.	Explore the functions of the eco-innovation ecosystem based on smart city blockchain	Smart and sustainable cities	Jiang and Zheng (2021)
Green technology innovation path based on blockchain algorithm.	Analyze the path of green technology innovation based on blockchain algorithms.	Development of green technologies	Li (2021)
How Industry 4.0 technologies and open innovation can improve green innovation performance?	Investigates the impact of Industry 4.0 technologies on green innovation performance and their mediating role of green innovation behaviour is studied.	Sustainability in industry	Mubarak et al. (2021)

Source: Elaborated by the authors.

Zhou et al. (2020); Hosow (2021); Li (2021); and Mubarak et al. (2021) point out in their papers that blockchain can have a negative effect on sustainability processes because the massive use of this technology can generate an increase in energy consumption and consequent CO_2 emission, and that studies about these impacts are necessary.

One aspect observed in common in these articles is that in applying blockchain for green innovation, there is the need for different actors such as government, universities, society, and companies. Thus, organizations are required to go beyond their existing knowledge base to explore new sources of knowledge - internally or externally - to overcome the challenges of this new technology. Two articles address this process within the innovation ecosystem perspective Hou et al. (2021) and Jiang and Zheng (2021). According to Mubarak et al. (2021), the degree to which green innovations can be enhanced through open innovation is a key policy debate to mitigate environmental issues. Authors such as Hosow (2021), Mubarak et al. (2021), and Zhou et al. (2021) discuss that there is a positive role of blockchain in enhancing open innovation, which leads to enhancing green innovation behavior. However, only in the article by Mubarak et al. (2021) is this relationship and proposition of improving the performance of "open green innovation" supported by digital technologies is discussed explicitly.

8. Conclusion

This article presents an exploratory study on research regarding the application of blockchain for open innovation processes and green innovation. The study is based on articles extracted from the SCOPUS database, Science Direct, and Web of Science.

Blockchain adoption for open innovation and green innovation processes has been incipiently occurring but already shows potential for expansion. The main contribution of these articles lies in the fact that studies focus on using blockchain to ensure digital trust and efficiency in digital collaboration for green open innovation processes. The analysis highlighted that the main research areas in the application of blockchain for green open innovation comprise the following areas: logistics and supply chain management, more sustainable processes and rational use of industrial resources, smart cities, energy efficiency, sustainable degrowth, smart agriculture, and maritime industry.

It is concluded that this field of research is still at an early stage and offers excellent scope and impetus for future academic work such as:

(a) drivers, barriers, and advantages for the adoption of blockchain and other digital technologies in large-scale green open innovation processes. This is mainly aimed at mitigating socio-environmental problems such as CO_2 emission decrease, smart and rational use of natural resources, energy efficiency, waste management;

(b) Change of business culture for implementation of green open innovation through blockchain;
(c) Studying the combination of blockchain with other technologies in the open green innovation process;
(d) Developing empirically tested frameworks for the use of blockchain in open green innovation processes;
(e) Applying blockchain for impacts on climate action and;
(f) applying blockchain to reward sustainable behaviors of society.

Blockchain technology is often considered one of the most defining innovations of the 21st century, and its application can generate remarkable change in all economic activities. When technological innovation must go hand in hand with environmental sustainability, it is essential to understand the relationships between blockchain and open, green innovation.

The aim of this paper is to points out gaps and trends for future research based on facts, and the literature reviews revealed that few studies have empirically explained performances by blockchain in the green open innovation processes, but only one paper deals with the topic explicitly (Mubarak et al., 2021).

Regarding the limitations of this paper, it is felt that the studies could have expanded the search of "strings" by analysis and comparisons with other digital technologies that used green, open innovation processes.

References

Aslan, D., Çetin, B.B. and Özbilgin, İ.G. (2019). An innovative technology: Augmented reality-based information systems. Procedia Computer Science, 158: 407–414.

Bai, Y., Song, S., Jiao, J. and Yang, R. (2019). The impacts of government R&D subsidies on green innovation: Evidence from Chinese energy-intensive firms. Journal of Cleaner Production, 233: 819–829.

Baldwin, C. and Von Hippel, E. (2011). Modeling a paradigm shift: From producer innovation to user and open, collaborative innovation. Organization Science, 22(6): 1399–1417.

Benítez-Martínez, F.L., Hurtado-Torres, M.V. and Romero-Frías, E. (2021). A neural blockchain for a tokenizable e-Participation model. Neurocomputing, 423: 703–712.

Blank, S. (2013). Why the lean startup changes everything. Harvard Business Review, 91(5): 64–72.

Bogers, M., Chesbrough, H. and Strand, R. (2020). Sustainable open innovation to address a grand challenge: Lessons from Carlsberg and the Green Fiber Bottle. British Food Journal.

Cheah, S.L.Y., Yuen-Ping, H.O. and Shiyu, L.I. (2020). How the effect of opportunity discovery on innovation outcome differs between DIY laboratories and public research institutes: The role of industry turbulence and knowledge generation in the case of Singapore. Technological Forecasting and Social Change, 160: 120250.

Chen, Y. (2018). Blockchain tokens and the potential democratization of entrepreneurship and innovation. Business Horizons, 61(4): 567–575.

Chesbrough, H.W. (2003). Open innovation: The new imperative for creating and profiting from technology. Harvard Business Press.

Chesbrough, H. (2006). Open business models: How to thrive in the innovation landscape. Harvard Business Press.

Chesbrough, H. and Bogers, M. (2014). Explicating open innovation: Clarifying an emerging paradigm for understanding innovation. New Frontiers in Open Innovation. Oxford: Oxford University Press, Forthcoming, 3–28.

Chesbrough, H. (2020). To recover faster from Covid-19, open up: Managerial implications from an open innovation perspective. Industrial Marketing Management, 88: 410–413.

Chen, Y. (2018). Blockchain tokens and the potential democratization of entrepreneurship and innovation. Business Horizons, 61(4): 567–575.

Dahlander, L. and Gann, D.M. (2010). How open is innovation? Research Policy, 39(6): 699–709.

Dahlander, L., Gann, D.M. and Wallin, M.W. (2021). How open is innovation? A retrospective and ideas forward. Research Policy, 50(4): 104218.

De Mergelina González-Santander, P. and Lemus-Aguilar, I. (2021). Current innovation sources driving the spanish electric power sector. Ingeniería e Investigación, 41(3).

Gassmann, O. and Enkel, E. (2004). Towards a theory of open innovation: Three core process archetypes.

Gattringer, R. and Wiener, M. (2020). Key factors in the startup phase of collaborative foresight. Technological Forecasting and Social Change, 153.

Hashimy, L., Treiblmaier, H. and Jain, G. (2021). Distributed ledger technology as a catalyst for open innovation adoption among small and medium-sized enterprises. The Journal of High Technology Management Research, 32(1): 100405.

Howson, P. (2019). Tackling climate change with blockchain. Nature Climate Change, 9(9): 644–645.

Howson, P. (2021). Distributed degrowth technology: Challenges for blockchain beyond the green economy. Ecological Economics, 184: 107020.

Jiang, Y. and Zheng, W. (2021). The coupling mechanism of green building industry innovation ecosystem based on blockchain smart city. Journal of Cleaner Production, 307: 126766.

Kassen, M. (2022). Blockchain and e-government innovation: Automation of public information processes. Information Systems, 103: 101862.

Katsamakas, E. and Xin, M. (2019). Open source adoption strategy. Electronic Commerce Research and Applications, 36: 100872.

Koh, L., Dolgui, A. and Sarkis, J. (2020). Blockchain in transport and logistics–paradigms and transitions.

Larsson, Z.Y., Di Gangi, P.M. and Teigland, R. (2019). Sharing my way to success: A case study on developing entrepreneurial ventures using social capital in an OSS community. Information and Organization, 29(1): 23–40.

Latusek-Jurczak, D. and Prystupa, K. (2014). Collaboration and trust-building in the open innovation community. Journal of Economics & Management, 17: 48–62.

Li, D. (2021). Green technology innovation path based on blockchain algorithm. Sustainable Computing: Informatics and Systems, 31: 100587.

Li, M., Li, Z., Huang, X. and Qu, T. (2021). A blockchain-based digital twin sharing platform for reconfigurable socialized manufacturing resource integration. International Journal of Production Economics, 240: 108223.

Li, Z., Wang, W.M., Liu, G., Liu, L., He, J. and Huang, G.Q. (2018). Toward open manufacturing: A cross-enterprise knowledge and services exchange framework based on blockchain and edge computing. Industrial Management & Data Systems.

Lichtenthaler, U. and Lichtenthaler, E. (2009). A capability-based framework for open innovation: Complementing absorptive capacity. Journal of Management Studies, 46(8): 1315–1338.

Lichtenthaler, U. (2011). Open innovation: Past research, current debates, and future directions. Academy of Management Perspectives, 25(1): 75–93.

Marshall, A., Dencik, J. and Singh, R.R. (2021). Open innovation: digital technology creates new opportunities—strategy & Leadership.

Massaro, M., Dumay, J. and Guthrie, J. (2016). On the shoulders of giants: Undertaking a structured literature review in accounting. Accounting, Auditing & Accountability Journal.

Mu, W., Bian, Y. and Zhao, J.L. (2019). The role of online Leadership in open, collaborative innovation: Evidence from blockchain open-source projects. Industrial Management & Data Systems.

Mubarak, M.F. and Petraite, M. (2020). Industry 4.0 technologies, digital trust, and technological orientation: What matters in open innovation? Technological Forecasting and Social Change, 161: 120332.

Mubarak, M.F., Tiwari, S., Petraite, M., Mubarik, M. and Raja Mohd Rasi, R.Z. (2021). How Industry 4.0 technologies and open innovation can improve green innovation performance? Management of Environmental Quality, 32(5, SI): 1007–.

Nakamoto, S. (2008). Bitcoin: A peer-to-peer electronic cash system. Decentralized Business Review, 21260.

Narayan, R. and Tidström, A. (2019). Circular economy inspired imaginaries for sustainable innovations. In Innovation for Sustainability (pp. 393–413). Palgrave Macmillan, Cham.

Nyame, G., Qin, Z., Obour Agyekum, K.O.B. and Sifah, E.B. (2020). An ECDSA approach to access control in knowledge management systems using blockchain. Information, 11(2): 111.

Ølnes, S. and Jansen, A. (2017, September). Blockchain technology as s support infrastructure in e-government. In International conference on electronic government (pp. 215–227). Springer, Cham.

Patrickson, B. (2021). What do blockchain technologies imply for digital creative industries?. Creativity and Innovation Management, 30(3): 585–595.

Park, G., Shin, S.R. and Choy, M. (2020). Early mover (dis)advantages and knowledge spillover effects on blockchain startups' funding and innovation performance. Journal of Business Research, 109: 64–75.

Passarelli, M., Ambrogio, G., Filice, L., Cariola, A. and Straffalaci, V. (2021). LiSC Model: An innovative paradigm for Liquid Supply Chain. Procedia Computer Science, 180: 893–902.

Pazaitis, A., De Filippi, P. and Kostakis, V. (2017). Blockchain and value systems in the sharing economy: The illustrative case of Backfeed. Technological Forecasting and Social Change, 125: 105–115.

Pazaitis, A., De Filippi, P. and Kostakis, V. (2017). Blockchain and value systems in the sharing economy: The illustrative case of Backfeed. Technological Forecasting and Social Change, 125: 105–115.

Pazaitis, A. (2020). Breaking the chains of open innovation: Post-Blockchain and the case of Sensorica. Information, 11(2): 104.

Podmetina, D., Soderquist, K.E., Petraite, M. and Teplov, R. (2018). Developing a competency model for open innovation: From the individual to the organizational level. Management Decision.

Prietula, M.J. (2014). Open collaboration for innovation: Principles and performance. Organization Science, 25(5): 1414–1433.

Rajput, S. and Singh, S.P. (2019). Connecting circular economy and industry 4.0. International Journal of Information Management, 49: 98–113.

Rahmanzadeh, S., Pishvaee, M.S. and Rasouli, M.R. (2020). Integrated innovative product design and supply chain tactical planning within a blockchain platform. International Journal of Production Research, 58(7): 2242–2262.

Robert, J., Kubler, S. and Ghatpande, S. (2020). Enhanced Lightning Network (off-chain)-based micropayment in IoT ecosystems. Future Generation Computer Systems, 112: 283–296.

Sandberg, M., Klockars, K. and Wilén, K. (2019). Green growth or degrowth? Assessing the normative justifications for environmental sustainability and economic growth through critical social theory. Journal of Cleaner Production, 206: 133–141.

Salampasis, D. and Mention, A.L. (2019). From a-value to value-multiplication: Leveraging outbound open innovation practices for unrelated diversification in the sensor industry. Technology Analysis & Strategic Management, 31(11): 1327–1340.

Spithoven, A., Clarysse, B. and Knockaert, M. (2010). Building absorptive capacity to organize open inbound innovation in traditional industries. Technovation, 30(2): 130–141.

Tang, K., Qiu, Y. and Zhou, D. (2020). Does command-and-control regulation promote green innovation performance? Evidence from China's industrial enterprises. Science of the Total Environment, 712: 136362.

Teodorescu, M. and Korchagina, E. (2021). Applying blockchain in the modern supply chain management: Its implication on open innovation. Journal of Open Innovation: Technology, Market, and Complexity.

Tranfield, D., Denyer, D. and Smart, P. (2003). Towards a methodology for developing evidence-informed management knowledge using systematic review. British Journal of Management, 14(3): 207–222.

Treiblmaier, H. and Sillaber, C. (2020). Um estudo de caso de transformação digital induzida por blockchain no setor público. Em casos de uso de tecnologia Blockchain e Distributed Ledger (pp. 227–244). Springer, Cham.

Vanhaverbeke, W. and Peeters, N. (2005). Embracing innovation as strategy: Corporate venturing, competence building, and corporate strategy making. Creativity and Innovation Management, 14(3): 246–257.

Vanhaverbeke, W., Van de Vrande, V. and Chesbrough, H. (2008). Understanding the advantages of open innovation practices in corporate venturing in real options. Creativity and Innovation Management, 17(4): 251–258.

Yang, X. and Li, W. (2020). A zero-knowledge-proof-based digital identity management scheme in the blockchain. Computers & Security, 99: 102050.

Yoo, Y. (2012). The tables have turned: How can the information systems field contribute to technology and innovation management research? Journal of the Association for Information Systems, 14(5): 4.

Yuan, G., Ye, Q. and Sun, Y. (2021). Financial innovation, information screening and industries' green innovation—Industry-level evidence from the OECD. Technological Forecasting and Social Change, 171: 120998.

Zhou, Y., Soh, Y.S., Loh, H.S. and Yuen, K.F. (2020). The key challenges and critical success factors of blockchain implementation: Policy implications for Singapore's maritime industry. Marine Policy, 122: 104265.

Zimmerling, E., Höllig, C.E., Sandner, P.G. and Welpe, I.M. (2019). Exploring the influence of common game elements on ideation output and motivation. Journal of Business Research, 94: 302–312.

CHAPTER 7

Blockchain and Sustainable Forest Supply Chain Management

A Literature Review

António Pimenta de Brito

1. Introduction

Blockchain technology was originally formed to power bitcoin, a digital currency. Its origins date back to the year 2008 when someone using the pseudonym of Satoshi Nakamoto, proposed to combine cryptology and an open ledger to create a digital currency application (Nakamoto, 2008). But blockchain is more than bitcoin. There are several blockchains, not just one. It's a distributed ledger of data, easily accessible to all people and is made to record and track anything of value, from a piece of art, to a money transaction or an electricity bill.

Why is it so special if there are already ways to track, store and process information? Why will blockchain revolutionize the way we interact with each other? It is primarily because of the way it tracks and stores data. Information is stored in units called "blocks", chronologically, in order to build a continuous line – a chain of blocks. If there is a change or update

ISCTE-Instituto Universitário de Lisboa/, ESCAD-IP, Luso – Universidade Lusófona.
Email: acpbo@iscte-iul.pt

in the information chain, the old information is not rewritten or erased, but one more block is created that stores the entire continuum of actions performed on that information. It is based on the accountant's model that records all financial movements in the ledger. It is a non-destructive way to store information changes over time (Center for International Governance Innovation (CIGI), 2018).

For example, when a person who owns a property sells it to someone else, this change is registered. Unlike the old method in which there could be a personal or organizational record of information, this is recorded in blockchain, which is decentralized and distributed over a large network of computers. This prevents the information from being tampered.

There are already many real-life applications of blockchain, such as in cryptocurrencies and fintech, but it is fairly new in the manufacturing sector (Iansiti and Lakhani, 2017). The connection of blockchain with supply chain (Saberi et al., 2018) is also fairly new, albeit exciting. It is this link between the blockchain and a sustainable supply chain that will be discussed in this study, specifically sustainable forestry sector supply chain.

Sustainable development is that which "ensures present goals without compromising those of future generations" (WCED, 1987, p. 43). There is a problem that makes sustainability of wood supply chains questionable, and that is illegal trafficking of the raw material. A traditional way of combating this phenomenon is through the certification of wood, but it is proven that this is a process that is easily violated and does not prevent the trafficking of wood globally. , It is estimated that 15 to 30 percent of all wood traded globally is illegal (Düdder and Ross, 2017). Blockchain can be a way to ensure that the wood used for processing has a legitimate origin.

In this study, a review of the academic literature on the applications of blockchain to sustainable supply chain management (SSCM) in the forest sector will be done. First, what the blockchain is, and what its business applications are, will be examined. Then, the concept of SCM and sustainable development will be presented. This study will be focused on the area of sustainable supply chain in the forest sector and the main problems it faces nowadays. Then the methodology used in this study, (namely, the conceptual framework of study) will be explained, after which the strategy and practice followed in the bibliographical research. Results of the literature review will be presented and, finally, conclusions will be drawn, limitations of this study underlined, and future research directions suggested.

The main contributions of this study are to the academic literature on Business/Management and as follows: it is the first and most recent literature review of the academic literature in the area of Business/

Management on the applications of the blockchain to the SSCM in the forest wood products sector. Finally, it is a useful introduction to topics such as Blockchain, Supply Chain Management, Sustainability and Industry 4.0.

2. Theoretical Framework

2.1 *Blockchain*

Blockchain technology creates confidence in information. For a block to be added to a chain, three things must happen:

1. A cryptographic puzzle must be solved, to create a block. The computer that solves this puzzle shares it among all the computers on the network (This is called "proof of work)".
2. It works without intermediaries: as there is information security, intermediaries such as banks and lawyers are dispensed with and thus, it can be safely transacted in a peer-to-peer manner, saving time and money.
3. As it is a technology and not a single network, it can be implemented in several ways. There are public, private and semi-private blockchains, as the case may be (CIGI, 2018). This is a technology that can revolutionize the world with the internet, but it poses a whole set of policy questions such as governance, international law, security and economics (CIGI, 2018).

An analogy that can help to understand the blockchain is, for example, the TCP/IP protocol (Iansiti and Lakhani, 2017), which revolutionized the internet. It was a disruptive innovation, foundational to the internet as the Blockchain is predicted to be. TCS/IPstands for "Transmission Control Protocol/Internet Protocol" (TCP/IP) and is a set of rules and procedures used to connect devices to the internet network. TCP/IP is also used as a communication protocol in private computer networks (intranet or extranet). TCP/IP specifies how communications should be made over the internet, by providing communications between users and identifying how information should be broken into packets, how it should be sent, transmitted, directed and received at the destination. The similarity to the blockchain is that TCP/IP does not need much centralized management and exists for networks to function reliably and to automatically recover if any device on the network fails (Shacklett et al., 2021). Similarly, when sending an email, the simple mail transfer protocol (SMTP) is the protocol that allows this operation and is a perfect, intermediary-free way to make it possible to establish an automatic peer-to-peer connection (Swan, 2017).

The digital money system like bitcoin, based on blockchain technology, allows to transfer money between one user and another automatically. With very low costs, in a few seconds, without the intermediation of central banks, instead of waiting for weeks or months, paying bank fees.

The four main forms of blockchain applications that are expected to develop in the future are: real-time money transfers and payments, property records, contract area and identity confirmations (Swan, 2017).

In an extended study to the existing academic literature on Blockchain, it was concluded that the most common use of Blockchain is in the area of finance and in cryptocurrencies, such as bitcoin. The most studied areas of blockchain are fintech, bitcoin and cryptocurrency. The most frequent keywords in the literature are: "blockchain", "bitcoin", "cryptocurrency", "fintech", and "smart contract". Five study clusters were found using the log-likelihood ratio (LLR) algorithm in Citespace: (1) "economic benefit"; (2) "blockchain technology"; (3) "initial coin offerings"; (4) "fintech revolution" and: (5) "sharing economy" (Xu et al., 2019).

Swan (2017) demonstrates the economic value of the blockchain through four applications, such as digital asset registries, leapfrog technology, long-tail personalized economic services, and payment channels a peer banking services:

a) Digital asset registries: A blockchain can be used to register, transfer and verify ownership of assets (be it a house, a work of art, a car or a building) and at the same time ensure that this data remains unaltered and protected from fraud (also applies to all personal records such as identity cards, passports, medical data, contracts, etc.).

b) Leapfrog technology: The leapfrog economy concept (Yayboke and Carter, 2020) was coined to demonstrate that a developing country can accelerate its progress by jumping phases of it to one that favors its growth, without the need to invest in intermediate phases. A common example is the availability of mobile phones and wireless network in underdeveloped countries. Thus, the investment in landline is bypassed and a financial and telecommunications revolution takes place. Through the use of this technology, several new technologies are promoted, access to mobile communications and the possibility of using mobile payments and credit cards are allowed. In the case of blockchain, due to the access, security, convenience and low cost of this technology, it could be a way to make mobile banking services accessible to thousands of people who do not have them. An example is the case of eWallet banking apps, a way to reach two million "unbanked" people in the world (PwC, 2016) or the case of China, with a high rate of use of mobile payments, could also be a suitable target of implementation.

c) Long-tail personalized economic services: In the old economy, companies spent big marketing budgets on those products that could have more acceptance and less investment in niche markets. In the long-tail economy, with the scale that the internet can offer to merchants and customers, there can be billions of niches on the internet that can be reached and can be highly profitable (Anderson, 2004). The 80/20 Pareto rule is replaced by the long tail economy. Blockchain can take advantage of this to offer the user endless possibilities of customizable services in an easy, secure and traceable way. Someone in the United States, in a few minutes, can buy a house in Ukraine using a decentralized marketplace owned store, or a customer can find an unfamiliar buyer on the Internet Marketplace and through a smart contract, perform an automatic transaction, without intermediaries, that is registered on the network, without the possibility of violation or compromise of the data.

d) Payment Channels and Peer Banking Services: There is a possibility of having person-to-person banking services, through a blockchain. They are payment channels for net transactions rather than gross payments. A blockchain can transact money like bitcoin, for example. Technology payment channels can evolve into a digital payment system.

However, blockchain is still evolving, and it would be necessary to change the entire communications paradigm, with regard to intermediaries and information privacy. To implement this technology, industries could start with smaller networks to test.

In the hypothetical future, with "smart contracts" with widespread implementation, financial and physical transactions can be carried out automatically, without the need for intermediaries such as lawyers, brokers and banks. The information would be transparent, automatic and traceable. A smart contract is one of the biggest blockchain opportunities. It is a program that allows one to automatically perform certain functions, according to certain pre-established conditions (Vilkov and Tian, 2019). This change would completely revolutionize how organizations and people perform and enforce contracts today. Assuming that requirements and assumptions are established in advance, shared and stored in the decentralized blockchain ledger, and shared across the entire network, here is no need for lawyers, banks or notaries, as all contracts are fulfilled automatically, with no scope for data tampering (destroying, manipulating or editing data) and intermediation by third parties.

However, when it comes to implementation, blockchain requires evolution, in terms of technology, governance and social impact. The two main threats to the implementation of blockchain are: (1) Difficulty and delay in harmonizing legal and banking systems as well as, consumer behaviour; (2) Security issues (Iansiti and Lakhani, 2017).

An experiment conducted in 2014 offered 4,494 MIT undergraduate students $100 each in bitcoin. Surprisingly, 30% of these students did not register to receive this free amount and 20% of those who registered converted bitcoin into cash after a few weeks. Even those who are used to the technology have had difficulties figuring out how to use bitcoin. Indeed, despite the hype this technology has created in the world of technology and finance, the most common criticism is the difficulty in changing well-established regulatory standards , influencing government policy and persuading central banks and large corporations to use this technology with broad acceptance. It can take years for the transition to take place (Iansiti and Lakhani, 2017).

The internet revolution has begun with a small set of experts as only they understood the intricate language of what they were creating. Then came others who made the technology intelligible to the common user; still the vast majority of users don't understand the architecture behind the common use of the internet. Blockchain is evolving similarly. Something that distinguishes the Internet revolution from the Industry 4.0 revolution is the widespread promotion of the need for literacy in programming languages for the common user. This was not so obvious before, when code makers and users were in perfectly delimited fields. Today, children learn programming at school, just as they would learn English earlier. This revolution, however, involves the coordination of efforts and agreements between many institutions about processes and standards. Political, regulatory, economic and social change can be enormous. Even in the case of smart contracts, the changes necessary for the implementation are plenty. New skills in blockchain software and programming have to be developed and mindsets have to change. Finally, several traditional sectors and jobs will have to disappear or be rethought. The expectation is that these applications will not reach widespread and critical mass adoption for another decade or probably more. One possible strategy to face the blockchain application is "starting small" (Iansiti and Lakhani, 2017) to develop know-how to then think bigger.

2.2 *The Supply Chain Management (SCM) Concept*

What was simply seen as logistics, has evolved into the supply chain concept. Both seek to improve operational efficiency and increase customer satisfaction, but supply chain management has a broader reach (Carvalho et al., 2020). In addition to including the activities of supply, transport, maintenance, storage, handling, manufacturing, distribution, it involves this need of collaboration between chain or channel partners, such as manufacturers, suppliers, intermediaries, service providers or customers (Council of Supply Chain Management Professionals (CSCMP),

2013). Therefore, the words collaboration, strategic integration, visibility and communication are essential in this context. Earlier there was more of an internal focus on a company and its sector (whether it was storage, manufacturing or transport); today the focus is external, and the keywords earlier mentioned are essential. From competition between companies, it moved to competition between supply chains (Christopher, 1992; Gold et al., 2009; Soler et al., 2010).

From the organizational efficiency, SCM has moved to concertation between business partners involved in the chain: customers, suppliers, R&D, logistics service providers, etc. From the retention of information, it has moved to the sharing of information among stakeholders. From the management of internal information, it has moved to a greater capacity for information integration and planning. There are several reasons for coordination between partners. One is to reduce inefficiencies, as information flows more effectively in the supply chain. Because there is more coordination between partners and more information flowing, there is much greater visibility of demand from production to end customer. If a retailer has the privileged information about the volume and quality of demand, it can also share this information with all stakeholders, this information is valuable in the suitability of the service, from the creation of a product to value creation at each stage of the supply chain. An obvious ultimate consequence of this collaboration between the partners is to reduce the cycle time of the chain. Simple logistics has evolved into supply chain management which today is a source of competitive advantage. Efficient management of the supply chain can provide cost advantage and product differentiation (Carvalho et al., 2020).

The primacy of information above all allows SCs to anticipate and respond to increasingly demanding customer preferences. Concepts such as "Just in time", associated initially with the automotive industry which companies as Toyota and Tesla use are based on the reduction of inventory and focus on customer responsiveness. This way, the accumulation of stock is avoided in the warehouse and the SC can respond to the customer quickly and customize the order to his preferences. This requires agile logistical and operational capacity and efficient information flow. but enables pply chain fulfills two of its main objectives: cost reduction and good customer service (Carvalho et al., 2020).

SCM affects all components of shareholder value creation: cost, customer service, return on assets and earnings (Beth et al., 2006). With the globalization of the economy, increased imports and exports, increased competitiveness and pressure on costs and the creation of global supply chains (World Bank, 2020), SCM has become necessary. Nowadays, a company can produce in one country, have research and development in another, storage and customization in another, have a distribution system and a network of retail stores spread across different parts of the globe, with multiple suppliers and in different time zones. The internationalization of companies and industrial relocation movements have seen exponential growth since the last quarter of the 20th century and consequently, productive specialization. With the development of a consumer society, products and services have become more complex and industries are witnessing increasingly rapid changes in the behavior of markets and the multiplication of segments.

There is a case to be made for product/service differentiation in customer service. There are increasing pressures to simultaneously improve service levels and reduce costs, due multiple reasons such as consumer activism, and enabling technologies such as the internet, artificial intelligence and industry 4.0. In this 4th Industrial Revolution, large amounts of transactions and data about consumption, operation and production are generated, stored and processed.

Supply chain disruption following the COVID-19 pandemic is a pressing today (Coupa, 2021; Gartner, 2021; Nandi et al., 2021). Supply constraints and the increase of raw material prices are raising questions about the viability of global supply chains. Going back to localization is also seen as a strong trend in Supply Chains (Nandi et al., 2021). Also, sustainability, the need for visibility, communication and digitalization are also strong issues to be consolidated in the coming years in supply chains (Coupa, 2021; Gartner, 2021). The need for information and collaboration along the chain has never been more important, in order to respond to demand, reduce costs and ensure a more resilient and agile supply chain. Blockchain is seen to be a solution to this problem and its implementation has been accelerated because of the pandemic (Gartner, 2021; Nandi et al., 2021).

A confrontational model in logistics with a large number of suppliers has evolved to a model of integration and collaboration with stakeholders of the SC in which information, visibility, transparency and traceability are essential.

One way of putting information sharing into practice throughout the supply chain is through technology. Systems such as ERP are widely used as an information interface along the supply chain, but they have several problems (Saberi et al., 2018). The purpose of using various technologies is to make possible and more efficient information, identification and traceability of products along the supply chain. Examples of these technologies include standard encoding systems such as EAN 13, EAN/UCC 128, EPC Global, automatic reading systems such as optical scanning, barcodes or RFID (Radio Frequency Identification). RFID is a technology that uses electromagnetic fields to automatically identify, and track tags associated with objects. This technology allows to connect real life objects with their virtual counterpart (Figorilli et al., 2018).

The internet of things (IoT) has also brought new dynamics to supply chain management. IoT is a term used to describe objects that communicate over the internet. In 1999, Kevin Ashton introduced the term "Internet of Things" to explain his concept of connecting radio-frequency identification (RFID) technology with the internet (Birkel and Hartmann, 2019). Is there a better way to fulfill the supply chain strategy, integration based on collaboration and information, visibility and traceability of tangible goods? This system facilitates the way information circulates on the internet. For better SCM IoT can be used to facilitate this sharing and information among network stakeholders (Xu, 2011). Haller et al. (2009, p. 15) defined IoT as: 'A world where physical objects are seamlessly integrated into the information network, and where they, the physical objects, can become active participants in business processes. Services are available to interact with these 'smart objects' over the Internet, query their state and any information associated with them, considering security and privacy issues'.

The data revolution and the placement of physical devices that can record and subsequently track, store and process information about the SC activity, opens up endless possibilities for visibility, transparency and improved efficiency and effectiveness throughout the supply chain. For example, a sensor present in an object can track its movement, temperature, perishability and many other indicators in real time and consequently store this information. This can be processed and treated for analysis and sharing. Predictive models and algorithms can be created and used based on this information.

The IoT technology applied to SCM opens up great opportunities in the efficiency, visibility and operability of SCM. However, there are still only a few analytical and empirical studies, or comprehensive applications across the entire supply chain. The best-known applications are in the agro-food sector, but often located in some parts of the value chain, not in an integrated manner. On the other hand, there are few IoT application models in an SCM environment and many barriers to implementation, namely technological and managerial. Information security and privacy is one of the greatest challenges (Ben-Daya et al., 2019; Birkel and Hartmann, 2019).

2.3 Supply Chain Management and Sustainability

The most cited definition of Sustainability is that of the World Commission on the Environment and Development (WCED), as presented in the Brundtland Report titled "Our Common Future". It defines sustainability as "development which meets the needs of the present without compromising the ability of future generations to meet their own needs" (WCED, 1987, p. 43). The United Nations has identified three areas as equally important pillars of sustainable development: economic, environmental and social (the Triple Bottom Line, TBL) (Sustainable Development Commission, 2011).

The issue of sustainability requires that the supply chain act in a way that manages to reconcile business, people and environmental objectives (Seuring et al., 2008) and allows the needs of stakeholders to be satisfied and the future of the planet guaranteed for generations to come.

A company participating in a global supply chain (World Bank, 2020) must be very attentive to the practices of its suppliers, in order to meet its objectives and those of its stakeholders. These practices have to do with socially and environmentally responsible and sustainable behaviors (Ashby et al., 2012). Therefore, in SCM's philosophy, sustainability cannot only come from one link in the chain, but that the chain is aligned in an integrated strategy, only in this way will it create value for the end customer.

Implementing ESG (Environment, Social and Governance) initiatives is a pressing issue in the coming years to Supply Chain Managers. COVID-19 appears to have worked as an accelerator of this trend (Coupa, 2021; Gartner, 2021; Nandi et al., 2021).

2.4 *Forest Wood Sustainable Supply Chain Management*

The forestry sector exists to commercialize wood and non-wood forest products. Wood products are produced using the timber of the trees (paper and furniture). Non-wood forest products are food and food additives (edible nuts, mushrooms, fruits, herbs, spices and condiments, aromatic plants), fibers (used in construction, furniture, clothing or utensils), resins, gums, and plant and animal products used for medicinal, cosmetic or cultural purposes (Food and Agriculture Organization of the United Nations (FAO), 2020).

The wood sector of forest products responsible for transforming the raw material – wood – into a final product such as paper, lumber, furniture and energy in various forms (heat, electricity and fuels) (Santos et al., 2019). Wood procurement and wood product industries are the areas that receive more research contributions worldwide, comparing to non-wood forest products. The forest wood products have multiple supply chains, but one thing in common, they are all based on one source, the wood from trees (Sjöström, 2000).

This study will deal with the supply chain of the forestry sector and its products as they are considered the primary forest products (FAO, 2020). In addition, it is the supply chain that figured the most in the bibliographical research, unlike non-wood forest products, with no known applications of blockchain to its activity. The concept of sustainability in the forest SCM has diversified since the triple bottom line became popular. Up until then it focused on one or two of the dimensions of sustainable development, mostly the economic and environmental. These concepts have been used in the literature simultaneously or in isolation, with the social pillar being the least studied or not studied at all (Santos et al., 2019).

According to a survey by Accenture (2018), forest industry leaders are very enthusiastic about the possibilities that blockchain can bring, with 60% of them stating that blockchain and smart contracts will be critical for organizations in the coming years and the first entrants are the ones who will dictate the rules of the game.

2.5 *Illegal Timber Traffic—The Main Problem*

The main problem encountered and addressed in the SSCM of forest products is illegal wood trafficking. 10 out of 13 of the papers analyzed in this study, and almost every source consulted, noted the problem.

Interpol (2019) declared that illegal logging accounted for 50 to 90 percent of all forestry sector activities in tropical forests and 15 to 30 percent of all wood traded globally (Düdder and Ross, 2017). The main sources of illegal activity in the forestry sector happens in Africa, Asia Pacific and Russia and Latin America (Seneca Creek Associates, LLC & Wood Resources International, LLC, 2004). 50 to 80 percent of forests from the Far East are cut down and exported to China illegally (Environment Investigation Agency (EIA), 2013). Southeast Asia, Russian Federation and Papua New Guinea are the top exporters of illegal timber, with shares of 55%, 20% and 11%, respectively (IUFRO 2016, cited by Vilkov and Tian 2019). Illegal trade of timber is highly profitable and estimated to be worth between USD 51–152 billion annually (Düdder and Ross, 2017).

In order to respond to this problem, environmental organizations and timber trading companies developed international standards to motivate the commercialization of wood from certified sustainable forests (Ozanne and Vlosky, 1997). However, certification - "chain of custody" (COC) - is ineffective; documentation related to legal timber trade is easily corrupted and good intentions come to nought (Düdder and Ross, 2017; Leipold, 2017).

Blockchain can be a way to ensure that when ordering raw material, it originates from reliable sources. Blockchain is a decentralized ledger that contains a chronological chain of data blocks, which are encrypted and store valid network activity. Just as this technology allows recording bitcoin transactions, it can also track physical assets. For example, a property, which is recorded for each asset (Abeyratne and Monfared, 2016). Therefore, it has been suggested as a way to track and record the ownership of natural products such as wood (Greenspan and Zehavi, 2016).

The country from which the wood comes is the strongest predictor of buyer confidence in forest products, secondly the price level of the product, thirdly certification, and finally the duration of the relationship with the supplier. With blockchain technology, it is possible to eliminate intermediation in the certification and purchase of raw materials and value-added products throughout the supply chain, accelerating the process and creating visibility and traceability throughout. By providing a history of movements within the blockchain, trust is created in the stakeholders involved and thus the sustainability of the product offered can be verified. It is proven that trust is a key element in the supplier-customer relationship, so the reliability conferred by certification and subsequent storage and availability using blockchain technology would realize this objective (Komdeur and Ingenbleek, 2021).

There is a huge problem in many countries due to the lack of transparency in relationships in the forest sector, leading to a high level of corruption and low level of cooperation in the supply chain. There is also an absence of a legal framework in the field of logging and timber trade and, finally few or no means of detecting and punishing this kind of offense, especially from Russia to China (Vilkov and Tian, 2019). The process of tagging and tracking the raw material and processing it in a blockchain network could act as a solution for a transparent and tamper proof supply chain (Vilkov and Tian, 2019).

There are already projects that use big data and blockchain to ensure safety and transparency in the forestry sector supply chain. For example, BVRio's Responsible Timber Exchange trading platform, created in Brazil, a country that deals heavily with the issue of illegal smuggling of timber, ensures that legal timber transactions are made with sustainable material, meeting the requirements of the European Union Timber Regulation and the US (Lacey Act, Costa et al., 2016, cited by Vilkov and Tian, 2019). However, even technology such as big data is vulnerable to corruption, centralized management and possibility of tampering and editing data (Vilkov and Tian, 2019).

Vilkov and Tian (2019) developed a SWOT analysis about the implementation of blockchain solutions in timber trading between Russia and China. Many of the conclusions in the model can be drawn for other countries, namely on the advantages and perils of Blockchain, because this technology is in its infancy (Table 1).

Table 1. SWOT analysis of the implementation of Blockchain in a SSCM of the forestry sector (Vilkov and Tian, 2019).

Strengths	**Weaknesses**	**Opportunities**	**Threats**
Impossibility of data substitution in the system of smart contracts	Complex technology and the early stage of development of example startups and pilot projects in the forest industry	Enterprise reindustrialization and transition to Industry 4.0 through IoT development	Implementation risks and reversability
Increased transparency, privacy and accurate tracking	Substantial financial and energy demands for blockchain-mining operations	Development of small and medium enterprises, and subsequent clustering	Immaturity of the Government – Science – Business model of regional and interregional scale
Elimination of illegal logging, corruption and fraud by absolute supply-chain sustainability	Complete loss of data on loss of ID	Creation of new jobs	Possible rejection by business of blockchain implementation in production service and international trade operations
Forest digitalization and increasing forestry sustainability	Insufficient or unsatisfactory functionality, security and regulation due to lack of standardization	Infrastructure modernization through timber supply chain development	
Compulsory forest products certification, improving the competitiveness of forest products, furniture, etc.	Competing platforms	Development of a digital economy	
Development of cryptocurrency mining	Reducing the number of jobs in forest products enterprises, and the displacement of all or some intermediaries, including banks		
Liquidation of intermediaries in logistics and banking (cost reduction)	The problem of international settlements on international timber trade transactions between Russia and China		
Distributed registry	Anonymity and open access to transaction information in case of its loss		

Evidence suggests that blockchain and IoT can be a good solution to make forest management more efficient. Technologies like 5G and LPWAN prove to be very promising, but the infrastructure needs, and entry costs make it a little impractical without the participation of other support industries willing to invest in this technology. Likewise, IoT is interesting but very expensive to implement, namely on a large scale, if large amounts of sensors are used (Henriques and Westerlund, 2020).

According to the interviewees in the study by Henriques and Westerlund (2020), Blockchain is a promising technology in the forest area, namely with the use of technologies such as the Hyperledger fabric and the R3 Corda. Hyperledger is linked to the implementation of the blockchain in large companies such as IBM and Walmart (Hyperledger Foundation, 2019). Blockchain and IoT are, thus, promising technologies when associated with the supply chain of the forest industry. However, these are still experimental applications (Henriques and Westerlund, 2020).

Figorilli et al. (2018) tested the first application of blockchain technology to wood tracking, from the georeferentiation of the standing tree to tags reading in the final consumer. Through RFID sensors, info tracing of wood is performed, from extraction, through the cutting, stacking, transportation, sawmill processing, to production and sale to the retailer and, finally, to the end consumer. These sensors ensure the traceability of the wood, since it is a tree, until it is cut and transformed into a final product. More technologies are used, such as a mobile phone application to process the information obtained by IoT, and the azure cloud in which the information is stored and processed. These help in the creation of the associated blockchain. This is a study of 15 logs in a forest in Italy, which resulted in 45 logs and in the end, tables bought by the consumer. The end customer can verify through a QR code or an RFID sensor, the provenance of the wood; and through a unique identifier from that piece, information about the evolution and transformation of the product till it reached him. This information is stored in the decentralized ledger of the associated blockchain, thus ensuring maximum security, data reliability and traceability throughout the entire wood supply chain.

By these experiments, it is proven that the use of blockchain can reduce up to 99% the illegal registration of wood in deforestation and illegal manipulation of wood areas (Cuéva-Sanchez et al., 2020).

There are blockchain decentralized apps used in the SSCM of the forest sector. The most popular mentioned in the literature for any supply chain, are: Ethereum DApp technology (Düdder and Ross, 2017, Sheng and Wicha, 2021), Hyperledger and R3 Corda (Düdder and Ross, 2017, Henriques and Westerlund, 2020).

According to Düdder and Ross (2017) a successful solution of blockchain applied to this sector has to fulfill the following requirements:

- Scalability for a large number of users, handling at least 100k transactions per minute.
- Efficient computation and verification of transactions for legality, environmental and social sustainability.
- Provision of a public key infrastructure for certifying transactions.
- Adaptability of the solution to mobile devices.
- Digital representation (data structure) of transaction data.
- Provision of physical proof of existence (for example, DNA trace) linked to a digital repository.

3. Research Methodology

The aim of this study is to understand the applications of blockchain technology in the forest SSCM, present in academic literature. The forest is part of a supply chain as explained above. and sustainability is important in its management.

The methodology followed in this study was to conduct a literature review on blockchain business applications in the forestry SSCM. First, a characterization and definition of blockchain and its main applications are presented. Then, an attempt is made to characterize what SCM is as it is relevant to the forest sector . The next step is approaching the forest SCM area within the context of sustainable development. In the end, it is intended to show what knowledge exists about blockchain applications to a SSCM of the forest sector. Finally, conclusions are generated based on the research findings.

This literature review seeks to answer the following research questions (Ashby et al., 2012):

1. What are the business applications of blockchain in the forestry SSCM found in academic literature?
2. What are the methodologies adopted in these studies and which are the dominant ones?

This type of literature review typically has three stages of: (1) Planning, in which the most relevant research questions are prepared; (2) Searching for relevant literature according to the criteria established in 1); (3) Drawing conclusions about what was found in the literature, namely the result of a content analysis and categorization. Finally, recommendations are made based on the conclusions drawn (see Figure 1).

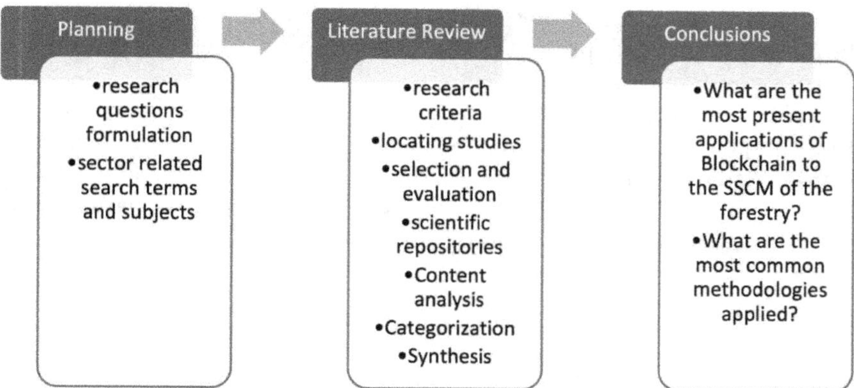

Figure 1. Research framework of the study.

Bearing in mind that the analysis is dealing with numerous areas and constructs - Blockchain, sustainability, SCM and forestry - all of them sufficiently broad concepts, several types of publications were considered. Otherwise, few results would be found, since the interconnection of all subjects significantly reduces the scope of investigation. Considering several types of publications allows the research to bring something new to science with this publication. However, the criterion that presided over the bibliographic research was that all studies should be peer reviewed. To reinforce the validity and replicability of this study this review focused specifically on blockchain applications in the forestry SSCM. and this literature review is completely unpublished in the Business/Management academic literature.

A typical methodological approach in paper reviews is to apply statistical sampling to a large number of results (Ashby et al., 2012; Burgess et al., 2006), but bibliographic research demonstrated that applications of blockchain in the SSCM forestry sector is still in its infancy and no review of this kind has yet been made in Business/Management literature with this actuality.

To review the literature on blockchain applications in sustainable forest SCM, a search was carried out in three of the most reliable repositories of academic literature that exist: Web of Science, Science Direct and Google Scholar.

Bibliographic search was carried out using three search terms in the aforementioned scientific repositories: "Blockchain forestry", "Blockchain timber" and "Blockchain wood". These terms were used as they represent the areas of forest sector that are in the scope of this study. In this context, there are terms such as "forestry" (sector), "timber" and "wood" related to the blockchain technology. The filters used in the main search were peer reviewed papers published since 2017, in English, with the two terms simultaneously present in the title of the publication, in the areas of Business and Management.

After selecting the sample resulting from the bibliographic search, these papers were read, and their content analyzed, and categorized. Content analysis and categorization will seek to answer the research questions previously mentioned. Finally, there will be a discussion about the findings that the research returned based on this analysis.

4. Results and Discussion

For the first term "Blockchain forestry" the following search filters were used: peer reviewed papers published since 2017, with the two terms simultaneously present in the title of the publication, in the areas of Business and Management. The Web of Science electronic database returned zero (0) results with this search, but 2 (two) results if the search is for publications under the terms "Blockchain timber" added to two knowledge categories: "Forestry" and "Materials Science Paper Wood". If the search is for "Blockchain wood" with the same filters, (not including knowledge categories) it returns two results, only one different from the previous search.

In the same search term, the ScienceDirect repository returned 17 (seventeen) results, but none of the terms "Blockchain" or "Forestry" were included in the title of the publication simultaneously. With a content analysis of the titles, it was also concluded that the topic of the publication was not about this theme, but mainly related to blockchain applications to SCM in general. Adding the category "Environmental Science" yields 71 results, but none of them contain both search terms in the title of the publication simultaneously. The search term "blockchain timber", with the same filters, gets 6 (six) results and none of them contain both terms in the title and subject of the papers. Finally, when searching for "blockchain wood" in Science Direct's search engine, it returns 86 (eighty-six) results, but none of them contain both the searched terms or topics covered in the title and subject.

When it comes to the Google Scholar search engine, it doesn't have the same filter refining capabilities as other electronic research sources, but it's not a repository that lags behind in searchability. The following filters were used: (1) peer reviewed papers published since 2017; (2) only review papers. The site returned 4170 publications. 14 of these publications were chosen, two of them already present in those found in the Web of Science. The search criterion was to choose publications that were related to Blockchain "Forestry", "timber" and "wood" in the title; results from up to page three of the respective search engine were chosen. It is concluded that Google Scholar has superior algorithmic capability as it returned results containing the search term "Blockchain Forestry" or not contained the search term in the title but was related with this subject. It returned results in which the papers were exactly about Blockchain applications in the SSCM area of the forest from a Business/Management perspective. The reason why this happens may not only be the algorithmic capacity of the search engine, but also the type of publications it returns in the results. It may be less demanding in the classification of publications according to scientific credibility.

A second conclusion that stands out in this study is that there is very little research on blockchain applications in the SSCM of the forestry from a Business/Management perspective.

From the literature review on the applications of blockchain technology to forestry, a categorization and classification by theme and methodology was carried out (see Table 2). It is possible to understand from the analysis of these works that the main problem facing the forest industry in terms of sustainability is illegal timber trade. Of the 13 publications analyzed, in 10 this problem is the main research question and blockchain is presented as a solution that can solve it. Other problems mentioned are the lack of transparency, security and efficiency in the supply chain, which can be mitigated by the use of blockchain technology. All of these issues are related to the previously described supply chain principles, which include the requirement for transparency, integration, traceability, efficiency, and trust in stakeholder relationships. The three major objectives of a sustainable supply chain that blockchain has the ability to achieve are cost reduction, customer happiness, and, lastly, ensuring the future of future generations in terms of business, people, and the environment.

Regarding the research methodologies used in the studies in question, 7 (seven) are theoretical studies and 6 (six) are empirical studies. Empirical studies are either tests with small samples and qualitative methodologies such as interviews, or the production of blockchain technology prototypes. The existing science in this area is promising but still emerging.

It is also verified that the two aspects of sustainability most discussed in this literature review are the environmental and the business ones. The social dimension is the least represented in literature in the SSCM area (Ashby et al., 2012) and in our sample only present in one study from the authors Willrich et al. (2020) (see Table 2). The reason why more results can be found in the business and environmental aspects is because the use of blockchain in this SC not only brings gains in terms of fighting wood trafficking but also introduces efficiency gains for the end customer and improvement of stakeholder relations. This way it strengthens the business/

Table 2. Results of the Literature Review: categorization and classification.

	Title of the Paper	**Type of Publication**	**Author**	**Problem Addressed**	**Methodology Applied**
1.	Blockchain as a solution to the problem of illegal timber trade between Russia and China- SWOT analysis	Scientific Paper	Vilkov and Tian, 2019	Illegal timber trading, traceability	Theoretical SWOT analysis
2.	Blockchain in forest products: Improving wood certification processes	White Paper	Accenture, 2019	Illegal timber trading, traceability	Theoretical paper
3.	Blockchain – killer of illegal Wood	Conference Paper	Lobovikov et al., 2018	Illegal timber trading in Russia	Theoretical paper of advantages and inconveniences of the use of Blockchain
4.	Blockchain-based Governance for Social Welfare in the Forestry	Conference Paper	Willrich et al., 2020	Lack of stakeholder data transparency and security in the forestry sector	Theoretical MCDA model
5.	Cryptocarbon: the promises and pitfalls of forest protection on a blockchain	Scientific Paper	Howson et al., 2019	Emissions from deforestation and forest degradation	Theoretical paper
6.	Digitalization of forest management: Next generation unsupervised monitoring using Internet of Things and Blockchain	Master's Thesis	Henriques and Westerlund, 2020	Forest inefficiencies, traceability of wood; transparency	Theoretical hypothesis of using Internet of things and blockchain to track and manage wood
7.	A Blockchain Implementation Prototype for the Electronic Open-Source Traceability of Wood along the Whole Supply Chain	Scientific Paper	Figorilli et al., 2018	Illegal timber trading; inefficiencies	Empirical study using RFID, IoT and open-source devices and tags

8.	How Blockchain Can Shape Sustainable Global Value Chains: An Evidence, Verifiability, and Enforceability (EVE) Framework	Scientific Paper	Nikolakis et al., 2018	Illegal timber trading, traceability, inefficiencies	Theoretical study based on a theoretical EVE framework model
9.	LogLog: A Blockchain Solution for Tracking and Certifying Wood Volumes	Conference Paper	Munoz et al., 2021	Illegal timber trading, traceability, inefficiencies	Empirical Study. Traceability system based on blockchain (prototype)
10.	A blockchain-based technological solution to ensure data transparency of the wood supply chain	Conference Paper	Cuéva-Sanchez et al., 2020	Illegal timber trading in Peru, traceability, inefficiencies	Empirical Study. Blockchain cloud-based system (prototype)
11.	The potential of blockchain technology in the procurement of sustainable timber products	Scientific Paper	Komdeura and Ingenbleek, 2021	Illegal timber trading, traceability, security, transparency	Empirical research about Blockchain technology
12.	The Proposed of a Smart Traceability System for Teak Supply Chain Based on Blockchain Technology	Conference Paper	Sheng and Wicha, 2021	Illegal timber trading in Thailand, traceability, transparency	Empirical Study. Blockchain technology traceability system based on Ethereum DApp
13.	Timber Tracking - Reducing Complexity of Due Diligence by using Blockchain Technology	Position Paper	Düdder and Ross, 2017	Illegal timber trading	Empirical Study. Blockchain based system to track timber (Prototype)

governance and sustainability aspect along with the environmental. These results can also be understood because most applications of blockchain technology are in business areas. "Cryptocurrencies", "fintech" and "bitcoin", are the main areas of research in the literature regarding blockchain applications (Xu et al., 2019). However, there are still a few case studies of blockchain applications in the forest sector SSCM (Howson et al. 2019).

5. Conclusions, Limitations and Future Directions

We live in an era where it is not just important to produce and consume. We realize that the planet's resources, the people that inhabit it and businesses are equally important. People extract and produce resources and provide services, but they must be in harmony with the other elements as a single system. Otherwise none of them will be here tomorrow. Sustainable development is an issue that is here to stay, as we know that resources are not endless, and the planet is at risk. People are also not disposable resources and have an inalienable dignity. Finally, there is no social and human development without companies and businesses, so they must be supported and made more efficient and effective.

Blockchain is a promising technology on several levels. It allows, on the one hand, efficiency and transparency; on the other, security and reliability of information. It has already demonstrated that it has applications in several areas; from digital finance to cryptocurrencies and smart contracts and there is growing investment. In the manufacturing field it continues to be underrepresented, but there are already many studies that prove that blockchain can revolutionize supply chains. With the outbreak of the COVID-19 pandemic, there is an increase in digitization and impact on logistics. Following these trends, blockchain technology implementation is likely to be accelerated. Also, sustainability is a trending topic and with be a strategic issue for years to come, according to studies consulted and the opinion of supply chain managers (Coupa, 2021; Gartner, 2021; Nandi et al., 2021). In this study, we analyzed the existing literature on blockchain applications in the forest wood products sector SSCM. We conclude that there is little academic literature on the subject, yet it is a promising area. We also conclude that the current most worrying problem that jeopardizes the sustainable development of forestry is illegal timber trade. Blockchain can be a solution to combat this scourge, to track and record the activity of the supply chain cycle from production phase, to supply, distribution and sale of wood, from raw material to final product, using RFID and complementary Industry 4.0 technologies, such as cloud computing and IoT. The most used methodologies in these studies were theoretical models and empirical studies, which included experiments with qualitative methodologies such as interviews and the production of prototypes of software systems based on blockchain. Finally, all publications deal with one or two of the following dimensions of sustainable development: environmental and business/governance, per the findings of Santos et al. (2019).

There are already some blockchain based business applications in SCM and the studies are promising, but more implementation and investigation are needed. Other factors that explain the slow evolution of blockchain applications for the forest SSCM can be high financial and energy costs, the lack of harmonization in terms of international practices and regulations, security issues and the need for a mindset change.

As limitations of this study, it is possible to point out the reduced number (N = 13) of studies consulted in the literature review. However, the area is very little studied (Howson et al., 2019). As future paths of investigation, more studies could be produced with prototypes that are tested in the real world by companies and using larger study samples. Secondly, more studies of blockchain applications in the following fields could be done; in non-forest products and on "social" aspects, the third pillar of sustainable forest development. It's a dimension little studied or not studied at all in the forest sector (Santos et al., 2019).

References

Abeyratne, S.A. and Monfared, R.P. (2016). Blockchain ready manufacturing supply chain using distributed ledger. International Journal of Research in Engineering and Technology, 5(9): 1–10.

Accenture. (2018). Technology Vision 2018: Intelligent Enterprise Unleashed. https://www.accenture.com/_acnmedia/accenture/next-gen-7/tech-vision-2018/pdf/accenture-techvision-2018-tech-trends-report.pdf.

Accenture. (2019). Blockchain in Forest Products: improving wood certification processes. Chemicals and Natural Resources Blog. https://www.accenture.com/us-en/blogs/chemicals-and-natural-resources-blog/blockchain-in-forest-products-improving-wood-certification-processes.

Anderson, C. (2004). The Long Tail. Wired. https://www.wired.com/2004/10/tail/.

Ashby, A., Leat, M. and Hudson-Smith, M. (2012). Making connections: A review of supply chain management and sustainability literature. Supply Chain Management: An International Journal, 17(5): 497–516.

Ben-Daya, M., Hassini, E. and Bahroun, Z. (2019). Internet of things and supply chain management: a literature review. International Journal of Production Research, 57(15-16): 4719–4742.

Beth, S., Burt, D.N., Copacino, W., Gopal, C., Lee, H.L., Lynch, R.P., Morris, S. and Kirby, J. (2006). Supply Chain Challenges: Building relationships. In Harvard Business Review on Supply Chain Management, 81(7): 64–73.

Birkel, H.S. and Hartmann, E. (2019). Impact of IoT challenges and risks for SCM. Supply Chain Management: An International Journal, 4(1): 39–61.

Burgess, K., Singh, P.J. and Koroglu, R. (2006). Supply Chain management: A structured literature review and implications for future research. International Journal of Operations & Production Management, 26(7): 703–729.

Carvalho, J.C., Guedes, A., Arantes, A.., Martins, A., Póvoa, A., Luís, C., Dias, J., Meneses, J., Ferreira, L., Carvalho, M., Oliveira, R., Azevedo, S. and Ramos, T. (2020). Logística e gestão da cadeia de abastecimento (3a Ed.). Edições Sílabo.

Center for International Governance Innovation (CIGI). (2018, Jan 14). What is Blockchain? The best explanation of blockchain technology (video). Youtube. https://www.youtube.com/watch?v=3xGLc-zz9cA.

Christopher, M. (1992). Logistics: The Strategic Issues (Ed.). Chapman & Hall.

Costa, P.M., Costa, M.M. and Barros, M. (2016). Using big data to detect illegality in the tropical timber sector. A case study of BVRio due diligence and risk assessment system. BVRio Institute. https://docspublicos.s3.amazonaws.com/madeira/BVRio-Big-data-to-detect-timber-illegality.pdf.

Council of Supply Chain Management Professionals (CSCMP). (2013, Aug.). Supply Chain Management Terms and Glossary. https://cscmp.org/CSCMP/Educate/SCM_Definitions_and_Glossary_of_Terms.aspx.

Coupa Software and WBR Insights. (2021, n.d.). Rethinking Risk and Opportunity in the Supply Chain. A survey of 200 CPOs on the year behind us and the year ahead. https://get.coupa.com/WBR-Survey-Report.html.

Cueva-Sánchez, J.J., Coyco-Ordemar, A.J. and Ugarte, W. (2020, Oct. 13–16). A blockchain-based technological solution to ensure data transparency of the wood supply chain. In IEEE (Org.). 2020 IEEE ANDESCON, Quito, Equador.

Düdder, B. and Ross, O. (2017). Timber tracking: Reducing complexity of due diligence by using blockchain technology. SSRN, 3015219.

Environment Investigation Agency (EIA). (2013, Oct.). Liquidating the Forests: Hardwood Flooring, Organized Crime, and the World's Last Siberian Tigers. https://content.eia-global.org/posts/documents/000/000/609/original/EIA_Liquidating_the_Forests.pdf?1479504214.

Food and Agriculture Organization of the United Nations (FAO). (2014). About Non-Wood Forest Products. [Online] Available at. http://www.fao.org/forestry/nwfp/6388/en/.

Food and Agriculture Organization of the United Nations (FAO). (2020, May). About Non-Wood Forest Products. http://www.fao.org/forestry/nwfp/6388/en/.

Figorilli, S., Antonucci, F., Costa, C., Pallottino, F., Raso, L., Castiglione, M., Pinci, E., Del Vecchio, D., Colle, G., Proto, A.R., Sperandio, G. and Menesatti, P. (2018). A blockchain implementation prototype for the electronic open-source traceability of wood along the whole supply chain. Sensors, 18(9): 3133.

Gartner. (2021, Nov.). Predicts 2022: Supply Chain Strategy. https://www.shippeo.com/en/resources/gartner-predict-2022-supply-chain-strategy.

Gold, S., Seuring, S. and Beske, P. (2009). Sustainable supply chain management and interorganizational resources: A literature review. Corporate Social Responsibility and Environmental Management, 17(4): 230–245.

Greenspan, G. and Zehavi, M. (2016, Jan.). Will provenance be the blockchain's break out use case in 2016? www.coindesk.com: https://www.coindesk.com/ provenance-blockchain-tech-app/.

Haller, S., Kanouskos, S. and Schroth, C. (2009). The Internet of Things in an enterprise context. In: Future Internet systems (FIS), 5468: 14–28.

Henriques, J. and Westerlund, W. (2020). Digitalization of forest management: Next generation of unsupervised monitoring using Internet of Things and Blockchain [Master of Science Thesis in the field of technology industrial engineering and management]. KTH Royal Institute of Technology – School of Industrial Engineering and Management Stockholm.

Howson, P., Oakes, S., Baynham-Herd, Z. and Swords, J. (2019). Cryptocarbon: The promises and pitfalls of forest protection on a blockchain. Geoforum, 100: 1–9.

Hyperledger Foundation. (2019). How Walmart brought unprecedented transparency to the food supply chain with Hyperledger Fabric. https://www.hyperledger.org/learn/publications/walmart-case-study.

Iansiti, M. and Lakhani, K.R. (2017). The truth about Blockchain—It will take years to transform business, but the journey begins now. Harvard Business Review. (Jan-Feb), pp. 118–127.

International Union of Forest Research Organizations (IUFRO). (2016). Illegal logging and related timber trade – dimensions, drivers, impacts and responses. A global scientific rapid response assessment report, (35).

Interpol. (2019, April). Global forestry enforcement. https://www.interpol.int/content/download/5149/file/Global%20Forestry%20Enforcement%20Prospectus%202019-web.pdf.

Komdeur, E.M. and Ingenbleek, P.T. (2021). The potential of blockchain technology in the procurement of sustainable timber products. International Wood Products Journal, 12(4): 249–257.

Leipold, S. (2017). How to move companies to source responsibly? German implementation of the European timber regulation between persuasion and coercion. Forest Policy and Economics, 82(82): 41–51.

Lobovikov, M., Pryadilina, N. and Scherbak, I. (2021, March 18–19). Blockchain–killer of illegal wood. In IOP Publishing . IOP Conference Series: Earth and Environmental Science, Saint Petersburg, Russian Federation.

Munoz, M.F., Zhang, K., Shahzad, A. and Ouhimmou, M. (2021, May). LogLog: A blockchain solution for tracking and certifying wood volumes. In IEEE (Org.). 2021 IEEE International Conference on Blockchain and Cryptocurrency (ICBC), Sidney, Australia.

Nakamoto, S. (2008). Bitcoin: A peer-to-peer electronic cash system. Decentralized Business Review, 21260.

Nandi, S., Sarkis, J., Hervani, A.A. and Helms, M.M. (2021). Redesigning supply chains using blockchain-enabled circular economy and COVID-19 experiences. Sustainable Production and Consumption, 27: 10–22.

Nikolakis, W., John, L. and Krishnan, H. (2018). How blockchain can shape sustainable global value chains: An evidence, verifiability, and enforceability (EVE) framework. Sustainability, 10(11): 3926.

Ozanne, L.K. and Vlosky, R.P. (1997). Willingness to pay for environmentally certified wood products: a consumer perspective. Forest Products Journal, 47(6): 39–48.

PwC. (2016). The Un(der)banked is FinTech's Largest Opportunity. DeNovo Q2 2016 FinTech ReCap and Funding ReView. https://www.strategyand.pwc.com/media/file/DeNovo-Quarterly- Q2-2016.pdf.

Saberi, S., Kouhizadeh, M., Sarkis, J. and Shen, L. (2019). Blockchain technology and its relationships to sustainable supply chain management. International Journal of Production Research, 57(7): 2117–2135.

Santos, A., Carvalho, A., Barbosa-Póvoa, A.P., Marques, A. and Amorim, P. (2019). Assessment and optimization of sustainable forest wood supply chains–A systematic literature review. Forest Policy and Economics, 10: 112–135.

Sheng, S.W. and Wicha, S. (2021, March 3–6). The proposed of a smart traceability system for teak supply chain based on blockchain technology. In IEEE (Org.). 2021 Joint International Conference on Digital Arts, Media and Technology with ECTI Northern Section Conference on Electrical, Electronics, Computer and Telecommunication Engineering, Cha-am, Thailand.

Seneca Creek Associates. (2004). Illegal Logging and Global Wood Markets: The Competitive Impacts on the US Wood Products Industry. American Forest & Paper Association.

Seuring, S., Sarkis, J., Müller, M. and Rao, P. (2008). Sustainability and supply chain management—An introduction to the special issue. Journal of Cleaner Production, 16(15): 1545–1551.

Shacklett, M.E., Novotny, A. and Gerwig, K. (2021, July). Definition – TCP/IP. https://www.techtarget.com/searchnetworking/definition/TCP-IP.

Sjöström, K. (2000). Logistics in the forest sector. Timber Logistics Club.

Soler, C., Bergstrom, K. and Shanahan, H. (2010). Green supply chains and the missing link between environmental information and practice. Business Strategy and the Environment, 19: 14–25.

Sustainable Development Commission. (2011). History of SD. http://www.sd-commission.org.uk/pages/history_sd.html.

Swan, M. (2017). Anticipating the economic benefits of Blockchain. Technology Innovation Management Review, 7(10): 6–13.

Vilkov, A. and Tian, G. (2019). Blockchain as a solution to the problem of illegal timber trade between Russia and China: SWOT analysis. International Forestry Review, 21(3): 385–400.

Von Carlowitz, H.C. (1713). Sylviceultura oeconomica. s.l.:s.n.

Willrich, S., Straub, T. and Weinhardt, C. (2020, Mar. 08–10). Blockchain-based Governance for Social Welfare in the Forestry. In (Org.). Wirtschaftsinformatik (Community Tracks). 15th International Conference on Wirtschaftsinformatik, Potsdam, Germany.

World Bank. (2020). Trading for Development in the age of Global Supply Chains. World Development Report. https://www.worldbank.org/en/publication/wdr2020.

World Comission on Environment and Development (WCED). (1987). Our Common Future, Oxford University Press, Oxford.

Xu, L.D. (2011). Information architecture for supply chain quality management. International Journal of Production Research, 49(1): 183–198.

Xu, M., Chen, X. and Kou, G. (2019). A systematic review of blockchain. Financial Innovation, 5(1): 1–14.

Yayboke, E. and Carter, W.A. (2020, April). The Need for a Leapfrog Strategy. https://www.csis.org/analysis/need-leapfrog-strategy.

CHAPTER 8

Blockchain as a Technology for Environmental Sustainability

A Bibliometric Review

Andreia de Bem Machado,[1,*] *Filippe Farias da Rocha*,[1] *Gertrudes Aparecida Dandolini*,[1] *João Artur de Souza*,[1] *Marc François Richter*[2] and *Marco Tulio Braga de Moraes*[1]

1. Introduction

The progress of organizations and society cannot occur in an irresponsible way with respect to social and environmental issues and it is necessary that society is aware of the historical period in which we live, with wide social inequality and environmental impacts never seen before with uncertain consequences.

In order for companies to achieve sustainable development for themselves and for their surroundings, it is essential that their collaborators have a critical vision and understand that the reality of the planet is the sum of the individual, organizational and governmental actions.

[1] Universidade Federal de Santa Catarina, Brazil.
[2] Universidade Estadual do Rio Grande do Sul, Brazil.
Emails: filipperocha@gmail.com; gertrudes.dandolini@ufsc.br; Joao.artur@ufsc.br; marc-richter@uergs.edu.br; marcotuliomoraes@gmail.com
* Corresponding author: andreiadebem@gmail.com

For this debate to be possible, the theme of social and environmental responsibility needs to be present in the training of professionals. With the worsening of ecological problems (global warming, desertification, silting, droughts, ...) and social crises (including humanitarian) occurring in various parts of the world, in 2015, Jorge Mario Bergoglio - Pope Francis, states in the "Encyclical Letter Laudato Sì: on the care of our common home" that " It's critical to look for holistic solutions that take into account natural systems' interactions with one another as well as with societal systems. There aren't two distinct crises: one environmental and one societal; rather, there is a single, complicated socio-environmental crisis." (2019). This systemic view of the ecological crisis and social inequality raises the understanding of socio-environmental responsibility to a level beyond the internal issues of the organization, its direct impacts on the environment or even the community of its manufacturing surroundings.

Society has been facing different challenges due to the numerous environmental impacts and socio-ecological problems experienced in the digital age that require innovative solutions. In this sense, Blockchain meets this new understanding, especially, about sharing knowledge in decentralized distributed networks that can be used for the sustainable development of our society. This technology has, among other possibilities, the ability to make relations between parties more secure, reliable and with fewer intermediaries. It is possible that humanity is facing a new socio-technical paradigm that will have an unprecedented influence on global society and will create opportunities for transformation in many sectors of the traditional economy. One example is the influence that blockchain is having on financial markets through cryptocurrencies, often known as "crypto-assets", with Bitcoin's exponential growth standing out.

Based on the foregoing, the major goal of this study is to map out how Blockchain technology might help with environmental sustainability through a literature assessment. To this effect, a bibliometric search was conducted using the Web of Science database. The research is divided, besides this introduction, into four parts. The first presents the concepts on Sustainability and Blockchain, the second addresses the methodology adopted in this research, the third part discusses the research results, and finally, the authors' final considerations.

2. Blockchain

The constant evolution of applications and refinement of concepts inherent to Blockchain certainly highlight the difficult mission of finding a consensus on the potential of this emerging technology. Likewise, the use of blockchain, until the end of 2021, present promising experiences that prove the benefits of this technology's use, especially in the solution of problems such as those encountered in the context of environmental sustainability.

From a pragmatic perspective, blockchain is a technology for recording and storing data in a distributed system (Di Pierro, 2017). Since its origin, with the publication of the paper "Bitcoin: A Peer-to-Peer Electronic Cash System" (Nakamoto, 2008), this technology is intended to establish trust between the parties in a system. In the technical-scientific literature, a blockchain is known by the acronym DLT - Distributed Ledger Technology and has among its attributes the quality of decentralization. According to the Organization for Economic Cooperation and Development (OECD), in the Working Papers on Public Governance, Blockchain technology is a type of distributed ledger technology that functions as an open and trusted record (i.e., a list) of transactions between two or more participants that is not held by a central authority (Berryhill et al., 2018).

In addition to allowing the registration of digital data, sharing them between the parties of a distributed system, the Blockchain also uses cryptography to secure transactions. Thus, a blockchain shares for all users of the network, identical copies of the records, with the respective chronological order, similar to a book of economic and financial activities (ledger), usually used in companies. Professor Andreas M. Antonopoulos, from the Department of Digital Innovation at the University of Nicosia/ Greece, highlights that a ledger in a blockchain can be understood in the same way as: "(...) a complete record of the economic activities of companies, generally used to track the transfer of money and ownership of assets" (Antonopoulos, 2017). For a better understanding about the structure of a blockchain, Figure 1 represents a ledger in a distributed system.

The immutability of activity records is another significant feature of blockchain, which allows the guarantee and effectiveness of transactions

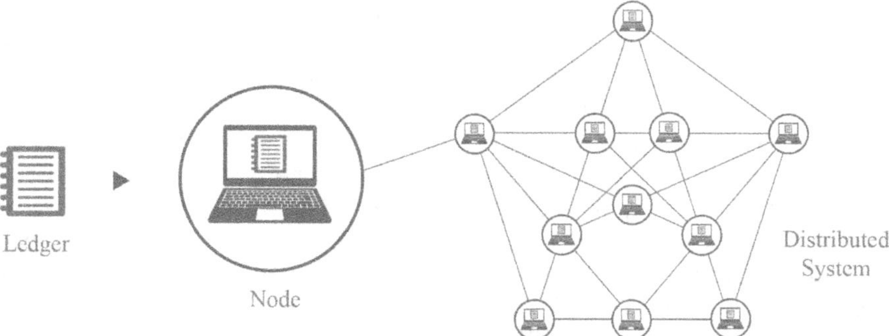

Figure 1. Illustrative image of a ledger in a distributed system. Source: the authors (2021).

in a system. In a blockchain environment there is the availability of verification and confirmation of authenticity of activities by anyone with access to the network, in order to ensure the visibility of the acts performed. As mentioned, blockchain technology was developed over the years in order to create a digital environment capable of recording transactions of value between parties, so that this ecosystem seeks to solve the existing trilemma between: security, scalability and decentralization.

Two technologies that have been consolidated for decades are key to enabling the efficiency of a blockchain. The first is cryptography, also known as cryptology, whose main function is the creation of codes to keep information secret. To protect communication between parties in the same community, data is converted into an unreadable format for unauthorized users and thus allowing the transmission of this data without unauthorized entities being able to decode it (Techopedia, 2011).

The second technology that enables blockchain to operate on global networks, such as the internet, is distributed computing. This refers to an area of computer science that studies distributed processing systems. A distributed system is one in which hardware or components on networked computers communicate and coordinate their operations only via the transmission of messages (Coulouris et al., 2013).

The innovation provided by blockchain happens from the integration and joint use of these two technologies. The result of this junction is a robust application, as it is extremely secure and scalable in order to store data capable of generating valuable information. Specifically, a blockchain is a form of immutable and chronological record, and at the same time, a technology that can be implemented for interactions between parties without a single central authority controlling the records of this information or data, thanks to the benefits of distributed computing.

Technological "avenues" can be represented through the implementation of the disruptive Blockchain network technology, incorporating features of a decentralized database enabling transaction on a global scale without intermediation or centralisation (Saberi, 2018).

It is important to note that there are public, hybrid and private blockchains, i.e., networks with different degrees of openness and permissions for users. The most emblematic example is the public bitcoin network, **recognized** as the most consolidated and comprehensive of the blockchain networks. At times, the public nature of the network is predominant, that is, with no restrictions, or permissions, for users to access it. Undoubtedly, the practical experience of the Bitcoin network shows that the public dimension of the blockchain is what ensures greater robustness in projects of this nature.

According to Swan (2015) the application of a blockchain is directly related to the governance capacity, to create innovative and specific solutions in networked organizations, that is, it enables the reconfiguration of governments, institutions, industries and, more broadly, most human activities. Blockchain governance enabled by information technology is a new, more efficient system for organizing, administering, coordinating, and recording all human interactions, whether in business, government, or personal life.

Blockchain definitions and concepts are still evolving, at the same time, recent studies point paths towards a consolidation of this new sociotechnical phenomenon. The World Economic Forum - Global Technology Governance Report 2021 highlights the role of blockchain as an option to the constraints imposed by the present global infrastructure, in response to crisis situations, such as the recent pandemic (COVID-19). As a "single source of truth", blockchain's immutability and transparency of transactions might boost trust in the veracity of key government data during a crisis (World Economic Forum, 2021).

In certain ways, the possibilities offered by blockchain technology have risen to prominence in recent years, especially those whose objective is to aid environmental sustainability. In this context, among the various applications of the technology the following stand out: the uses of blockchain for integrating sustainable energy systems (Wu, 2018), creation of circular economy (Kouhizadeh, 2019), sustainability of supply chains (Yadav, 2019) and, more recently, initiatives to mitigate the impacts of pandemic (COVID-19) in the management of global supply chains (Bhaskar et al., 2020). Table 1 presents some applications of blockchain in the context of environmental sustainability, according to the previously stated writers' understanding.

Table 1. Applications of blockchain technology in the context of environmental sustainability.

Blockchain Technology Applications
• Peer-to-peer (P2P) Energy Transaction • Physical Information Security • Carbon Emissions Certification and Trading • Virtual Power Plant • The Synergy of the Multi-Energy System • Information and communication Technologies (ICTs) in agriculture, forestry and fishing industry • Mining industry: exploration, reserve estimation, mine design and planning processes • Manufacturing in the industry 4.0 era • Transportation, communication and utilities services • Retail trade • Supply chain network

Source: Wu (2018); Kouhizadeh (2019); Yadav (2019); Bhaskar et al. (2020).

At this time the world is facing issues related to sustainability (Saberi, 2018). Sustainability has many definitions. ranging from an intergenerational philosophical approach, that is concerned that future generations are not negatively affected by decisions taken today, to a multidimensional approach, characterized as the balance of environmental, social and business dimensions (Seuring et al., 2008).

Blockchain technology can contribute to the reduction of unethical, corrupt and counterfeit practices, due to its transparency, traceability and immutability; this may be one of the main contributions of this technology regarding sustainability (Kouhizadeh et al., 2020).

There is no doubt that contemporary society, increasingly interdependent, globalized and composed of complex relationships, will undergo profound changes in the coming years. Blockchain technology should not be understood as a unique and effective solution, but as a powerful technology that will allow transformations in various industry sectors and other activities of society, especially in extremely complex situations such as the current environmental problems that plague humanity.

3. Environmental Sustainability

Environmental sustainability is defined as "the responsible use of natural resources" in order to ensure that these resources will also be available to future generations. According to Mancebo and Sachs (2015), environmental sustainability refers to the capacity of ecosystems to sustain, absorb and recompose. The authors state that "environmental sustainability can be achieved by intensifying the use of potential resources for socially worthwhile purposes; limiting the consumption of fossil fuels and other easily exhaustible or environmentally harmful resources and products by replacing them with renewable and/or abundant and environmentally harmless resources or products; reducing the volume of waste and pollution as a whole; and intensifying research into clean technologies." The concern with sustainability is fundamental to reduce environmental problems such as: pollution, greenhouse effect, global warming, extinction of animals and plants and the end of natural resources. One of the biggest challenges in applying environmental sustainability measures is to find a balance between the economic and social development of a country and the preservation of its environment (Thangavel and Sridevi, 2016).

In the 1970s, there was a growing concern about environmental sustainability, when the need for environmental preservation began to draw attention. The perception of the damages that resulted from the increase in industrialization, which started in the 19th century with the

Industrial Revolution, was determinant for the understanding of this need. But only in the 1980s, UN World Commission on Environment and Development coined the term and suggested the first measures of environmental sustainability (Machado and Richter, 2020).

It is important to understand the concept of "sustainability" in order to understand environmental sustainability. Sustainability is based on the union of the three most affected areas: environment, society and economy. For sustainability to become a reality as a whole, these three areas must be considered together, inseparably, and the measures created must encompass all these interests. In other words: environmental sustainability is one of the three pillars of sustainability (Purvis et al., 2019).

The concern of companies with transparency in the compliance of environmental governance principles has created an opportunity for blockchain. The technology has been integrated into the business of companies with the objective of monitoring actions aimed at the preservation of the environment and ESG (Environmental, Social and Governance) practices. With blockchain it is possible not only to register information about the movement of the corporate sector towards good sustainability practices, but also to 'supervise' how the responsibility in relation to the sustainability issue is progressing. The use of blockchain creates an environment of greater trust between the actors involved in the different production chains due to its characteristics. Blockchain has greater cybersecurity by creating immutable audit trails, with transparency and secure information sharing (Hull et al., 2021).

The agricultural and livestock sector has had a special highlight in the use of blockchain, for the purpose of tracking the animal protein chain and contributing to the reduction of CO_2 emissions. Thus, it could be verified that, in 2019, beef cattle ranching accounted for 62% of Greenhouse Gas (GHG) emissions, in Brazil, according to the Climate Observatory's GHG Emissions Estimates System (IPCC, 2019; Cusack et al., 2021).

On the other side, blockchain could prevent the destruction of forests, such as the Amazon rainforest in the northern region of Brazil. The proposal of two enthusiasts of the use of 4.0 technologies for sustainability, the environmentalist brothers and Brazilian researchers Carlos Nobre and Ismael Nobre, is for Brazil "to become a bioeconomy combining the knowledge of our biodiversity with the possibilities of Industry 4.0". As the industrial sector is increasingly interested in what the Amazon has to offer for its production (such as raw materials for medicines and cosmetics), it is time to take advantage of this in the right way and with gains for the communities in the region. They created the Amazônia 4.0 project which, among various actions, includes the Amazon biobank. The use of blockchain by the biobank began in 2022. The genomic sequencing of several forest species will be stored there. Thus, it will be possible to have a permanent and secure record and control of where the information comes from and where it goes to, guaranteeing the intellectual property of the data, thus reducing the problem of biopiracy (Nobre and Nobre, 2020).

In this way, Brazil has much to do and can become a global example of an economy that combines, in a "friendly" way, business, technology and the environment. The European Union (EU) is also going down this bioeconomy path. The bloc launched the European Green Deal at the end of 2019. On the one side, it's about lowering CO_2 emissions, but it's also about encouraging job creation and creativity, an example of the unification of the three pillars of sustainability to deal with issues such as biodiversity loss, sustainable agriculture, imposing environmental rules and penalties for the polluting.

As the Fourth Industrial Revolution picks up steam, inventions and innovations are becoming more efficient, quicker, more broadly available than ever before. We are now seeing a confluence of the digital, physical, and biological domains as technology becomes more integrated. Emerging technologies such as the Internet of Things (IoT), virtual reality, and artificial intelligence (AI) are allowing societal changes that will have far-reaching implications for future generations' economies, values, identities, and opportunities. There is a unique opportunity to harness the Fourth Industrial Revolution, and the social changes it triggers, to assist in the resolution of environmental concerns and the transformation of how we manage our common global environment (Herweijer et al., 2018; Schwab and Davis, 2018).

Seizing these opportunities and proactively managing these risks will require a transformation of the approach to global environmental management. This covers governance structures, procedures, investment and funding methods, the prevailing incentives for technical advancement and the current form of social engagement. This transition will not occur by itself; it will need deliberate coordination among policymakers, scientists, civil society leaders, technology advocates, and investors.

Below are some areas and opportunities where Blockchain technology can (once the aforementioned challenges are overcome) be the baseline and add immense value in the pursuit and realisation of a more sustainable development for the planet (Schwab and Davis, 2018).

- New sources of sustainable finance: Blockchain-enabled financial platforms can potentially revolutionise access to capital and unlock the potential for new investors in projects that address environmental challenges; from investing in "green" infrastructure projects to facilitating structured finance transactions or humanitarian donations to developing countries (Schletz et al., 2020). On a broader level, there is the potential for Blockchain to facilitate a shift from a system more focused on shareholder value to a system with value for stakeholders as well, and also to expand traditional financial capital accounting so that it can capture social and environmental capital. These reforms, taken together, might help raise the trillions of dollars required to fund the transition to low-carbon, sustainable economies.

- Encouraging circular economies: The value and exchange of materials and natural resources might be profoundly altered by blockchain, encouraging individuals, businesses and governments to recover value from items that are now thrown, squandered, or considered economically worthless. This could drive widespread behavior change and help realize a true circular economy (Rotabi and Ali, 2022).
- Transforming the carbon (and other environmental) market: Blockchain platforms might be used to optimize existing carbon (or other hazardous material) market platforms and generate new opportunities for carbon credit trades by using cryptographic tokens with transferable value (Schletz et al., 2020).
- Sustainability monitoring, reporting and verification: Blockchain has the potential to revolutionize sustainability reporting and assurance, allowing businesses to better monitor, show, and improve their results, while allowing consumers and investors better information for decision making. This could drive a new wave of accountability and action as this information can more effectively reach the senior management level of organizations providing them with a more complete picture to manage risk and reward profiles.
- Automated disaster preparedness and relief: Blockchain could support a new shared system connecting numerous groups interested in disaster relief and preparedness in order to improve efficiency, effectiveness, coordination, and accountability of entrusted resources. A decentralized interoperable system could enable information sharing (e.g., activities transparent to all other parties within the network) and rapid automated transactions through contracts. This might increase efficiency in the aftermath of disasters, which is the most essential moment for limiting loss of life and other human consequences (Ozdemir et al., 2020).

4. Methodology

With the objective of increasing knowledge, measuring and analyzing scientific literature publications on the subject matter of the "Trust that blockchain technology build to assist in environmental sustainability", a bibliometric analysis was conducted. This was done with a search of the Clarivate Analytics' Web of Science (WoS) database. The study was developed using a strategy consisting of three phases: execution plan, data collection and bibliometrics. To analyze the bibliometric data, the bibliometrix software was used because it is the most compatible with the Web of Science database. The R Bibliometrix package, called Biblioshiny, has the most extensive and appropriate set of techniques among the tools surveyed for bibliometric analysis (Moral-Muñoz et al., 2020). In a bibliometric study, these statistics organize relevant information such as chronological distribution, main authors, institutions, and countries, kind of publishing in the field, and main key words and the most referenced papers. From a statistical standpoint, scientific mapping allows researchers to investigate and construct a worldwide picture of scientific knowledge.

4.1 Methodological Approach

The study to address the research problem stated, is classified as exploratory-descriptive in order to describe the theme and increase the familiarity of researchers with facts. A systematic search in an online database was employed for the literature search, which was followed by a bibliometric analysis of the results. Bibliometrics is a method used in the information sciences to map texts based on bibliographic records maintained in databases using mathematical and statistical methodologies (Linnenluecke et al., 2020). Bibliometrics allows relevant calculations such as: number of productions per region; temporality of publications; organization of research by area of knowledge; quantity of literature relating to the study's citation; identification of a scientific publication's impact factor, among other contributing factors to the systematization of research findings and the minimization of the occurrence of biases when analyzing a given topic. For the bibliometric analysis, the study was organized into three distinct stages: planning, collection and results. These stages took place in a convergent manner to answer the study's guiding question, namely: How can Blockchain technology may support environmental sustainability?

Planning started in the month of September 2021, and ended in the month of December 2021, when the research was conducted. During this phase, various criteria were established, such as limiting the search to digital databases and excluding physical library catalogues, due to the number of documents deemed sufficient in the research bases in the database chosen in this research. The scope of the study was focused on the WoS database due to its relevance in the academic field, and its interdisciplinary character. WoS was also chosen because it is one of the largest databases of abstracts and bibliographic references of peer-reviewed scientific literature and because it is continually updated. In the planning phase, search terms relating to the research namely "Blockchain" and "Sustainability" and "Technology, were defined.

Variations of the expressions used for the search are presented, in a larger context, within the same proposal, because a concept depends on the context to which it is related. Finally, when planning the search, it was decided to use the terms defined in the "title", "abstract" and "keyword" fields, without making temporal, language or any other restrictions that might limit the results.

Considering the research problem: search terms were delimited in the planning phase. First the following descriptor was chosen: How can Blockchain technology assist in environmental sustainability? In order to refine the search aligned to the research problem, an additional search was conducted with the following terms: "Blockchain" and "Sustainability" and "Technology" which originated 445 documents found.

In research planning phase, the data collection retrieved a total of 445 indexed papers. The first publication was in 2016. For this study, papers published between the years 2016 and 2021 were considered.

4.2 Bibliometrics Results

As a result of this collection, it was found that the 445 papers identified in the search were written by 1372 authors, linked to 785 institutions from 65 different countries. A total of 1509 keywords were used. Table 2 illustrates the outcomes of this bibliometric analysis data collection.

The publications in the Web of Science database that are eligible were published between 2016 and 2021. In the year 2016, there were three publications. In the year 2017, four papers were published. In the year 2018 a total number of 22 publications were retrieved. And in the years 2019, 2020 and 2021 there was a significant increase in publications, with 62 articles in 2019, whereas in the years 2020 and 2021 there were, respectively, 150 and 204 publications, according to Figure 2.

From the 445 papers a varied list of authors, institutions and countries is observed, that stands out in the research on how blockchain technology can assist in environmental sustainability. When analyzing the 20 countries with the highest number of citations in the area, it can be seen that the USA stands out with an average of 28% of a total of 1442 citations, while China, in second place, stands out with 14%, according to Figure 3.

Figure 4, presents the intensity of publication by country and the relationship established between them, through citations between published papers.

Table 2. Bibliometric data.

Description	**Results**
MAIN INFORMATION ABOUT DATA	
Timespan	2016:2021
Sources (Journals, Books, etc.,)	197
Documents	445
Average citations per document	12,07
Average citations per year per document	5,948
References	23229
DOCUMENT TYPES	
Article	297
Article; early access	31
Article; proceedings paper	1
Editorial material	4
Editorial material; early access	1
Proceedings paper	53
Review	53
Review; early access	5
DOCUMENT CONTENTS	
Keywords Plus (ID)	754
Author's Keywords (DE)	1509
AUTHORS	
Authors	1372
Author Appearances	1552
Authors of single-authored documents	24
Authors of multi-authored documents	1348
AUTHORS COLLABORATION	
Single-authored documents	24
Documents per Author	0,324
Authors per Document	3,08
Co-Authors per Document	3,49
Collaboration Index	3,2
Affiliations	785

Another analysis performed is related to the identification of authors. The authors who have more production in the area are: J. Sarkis with 13 articles, M. Kouhizadeh with six, and B. Wang and X. Wang with five publications, each, in the area, as shown in Figure 5.

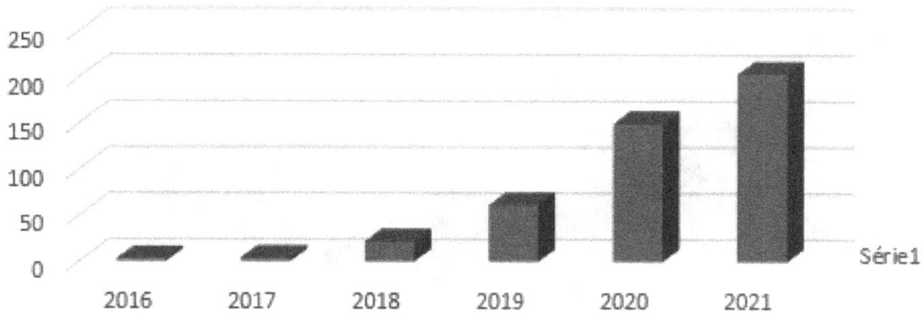

Figure 2. Temporal distribution of publications. Source: The authors (2021).

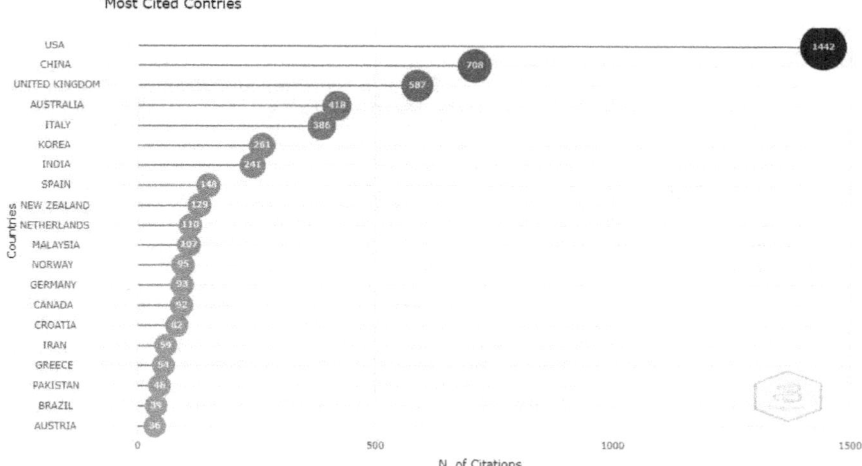

Figure 3. Distribution of work by country. Source: The authors (2021).

Authors that were most cited globally: J. Sarkis, with 142 citations, M. Kouhizadeh with 135, S. Saberi with 115 and L. Shen with 97 citations, as illustrated in Figure 6.

The productivity and relevance of the production over time of the main authors are represented in Figure 7, where the size of the point represents the number of publications and the intensity of the color is the number of citations of the annual publications. It is observed that the productivity and the relevance of publications, over time, are dynamic.

The top ten research areas publishing on this blockchain, technology and sustainability topic are: Ecology Environmental Sciences, Other Science &Technology Topics, Computer Science, Engineering, Business Economics, Operations Research, Management Science, Telecommunications, Energy Fuels, Food Science Technology, Information Science and Library Science.

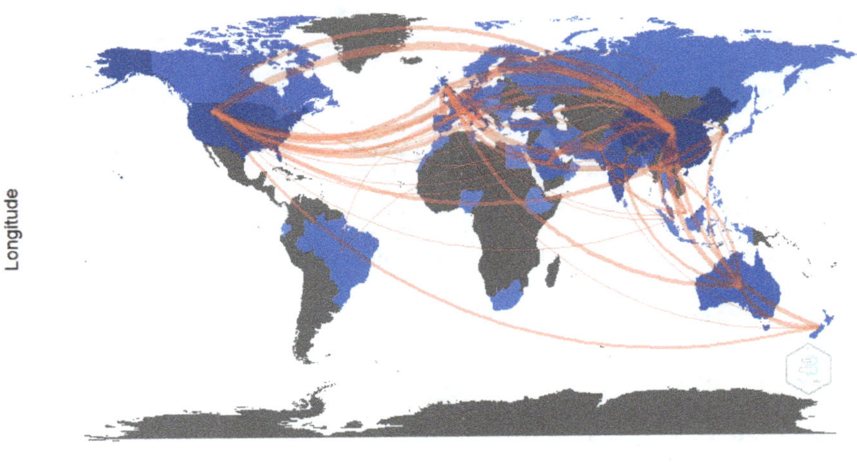

Figure 4. Spatial distribution and relationships of publications on Blockchain, sustainability and Technology. Source: The authors (2021).

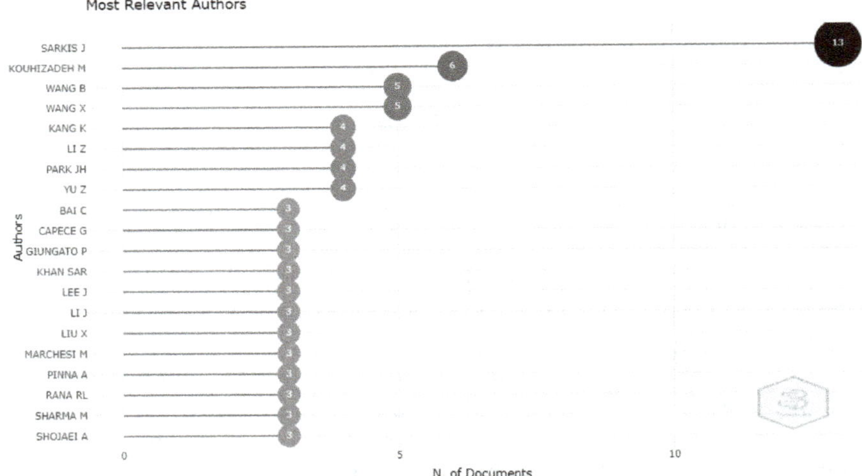

Figure 5. Authors with the highest number of publications on the subject of the search. Source: The authors (2021).

With 27% of the publications, the areas of Ecology Environmental Sciences and Other Science Technology Topics stand out, followed by Computer Science with 12%.

The documents analyzed were published in 197 different journals, and among the total of 445 studies, 152 (35%) were published in a

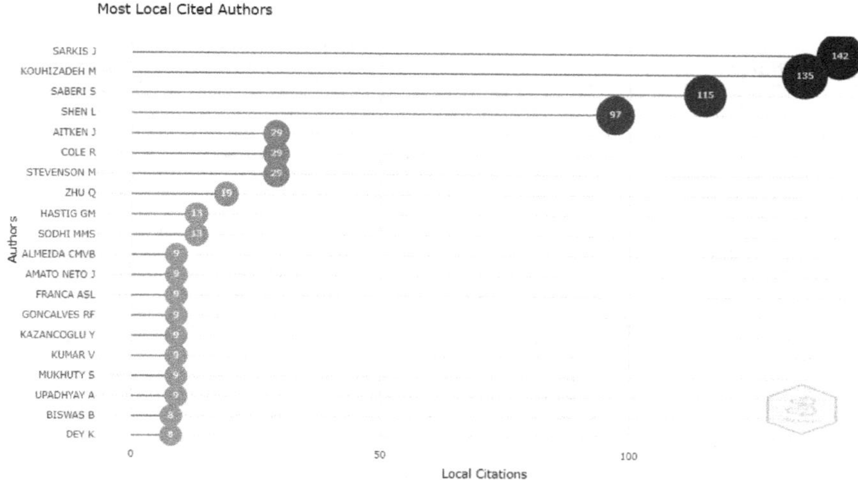

Figure 6. Authors with the most citations worldwide. Source: The authors (2021).

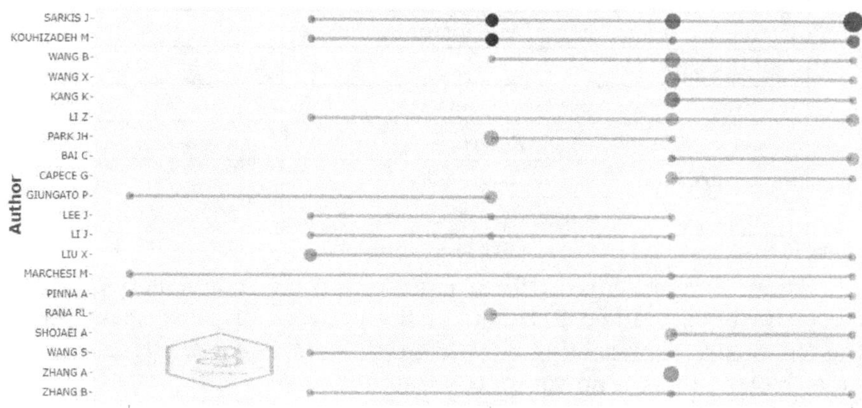

Figure 7. Productivity and relevance of publications by author. Source: The authors (2021).

single journal Sustainability, according to Table 3, which presents the 20 scientific sources with the highest number of publications on the subject "Blockchain, technology and sustainability".

The second most relevant source, among the twenty explained in Table 3, is the "Journal of Cleaner Production" with 17 documents. Next in sequence are, with eight publications, the "IEEE Access" and "International Journal of Production Research", followed by the journal "Frontiers in Blockchain" with seven papers.

Table 3. Top scientific sources with most publications.

Source (Journal Title)	Number of Articles
Sustainability	152
Journal of Cleaner Production	17
IEEE Access	8
International Journal of Production Research	8
Frontiers in Blockchain	7
Annals of Operations Research	5
Energy Research & Social Science	5
Entrepreneurship and Sustainability Issues	5
International Journal of Information Management	5
Trends In Food Science & Technology	4
Business Strategy and The Environment	3
Computers & Industrial Engineering	3
Computers in Industry	3
Foods	3
Information Technology for Development	3
Logistics-Basel	3
Renewable & Sustainable Energy Reviews	3
Resources Conservation and Recycling	3
Sustainable Production and Consumption	3
Technological Forecasting and Social Change	3

Figure 8 represents in three clusters the co-citation network. The cluster in blue, marked S. Saberi, is the network of greatest influence; the red cluster, representing S. Nakamoto, comes second; whereas in the green cluster, there is no author of great influence.

Figure 9 demonstrates the collaboration network of 20 "nodes each of which symbolizes each author, with the edges indicating the coauthorship links between them. The cluster in blue, indicating S. Saberi's work, has the network of greatest influence, and in second place N. Kshetri. In the red cluster, S. Nakamoto stands out considerably from the others.

Figure 10 represents an innovative three-field graph, in which, in the observed columns, from left to right, the interactions between, author keyword, authors and most relevant countries are shown.

As of the general search, it was also possible to analyze the hierarchy of research sub-branches in the blockchain, technology and sustainability area. The set of rectangles represented in the TreeMap, where the words that stand out are "Blockchain" with 273 occurrences, "Sustainability"

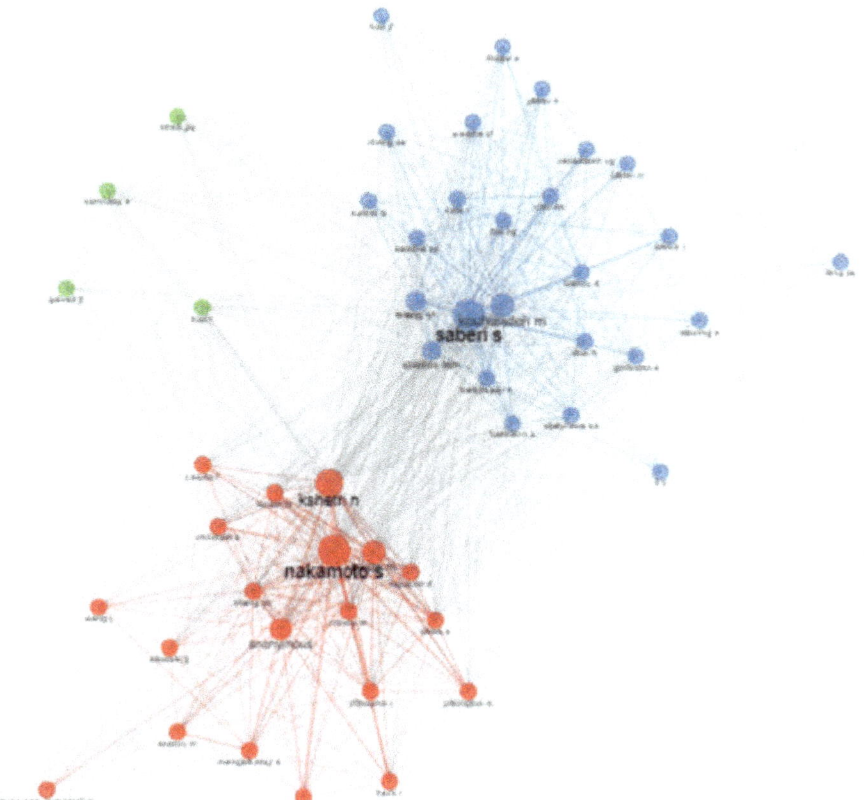

Figure 8. Authors' co-citation network. Source: The authors (2021).

with 128 and "Blockchain technology" with 50 occurrences. Figure 11 shows in a proportional manner, the hierarchy of research sub-branches.

As of the bibliometric analysis, based on the group of papers retrieved, and 1509 keywords indicated by the authors, "Blockchain" stood out with 273, "Sustainability" with 128 and "Blockchain technology" with 50 occurrences, as per Figure 12. "Supply chain" also stands out, which leads to the conclusion that this is an area of application of blockchain technology that has been widely explored in scientific literature.

The twenty most cited documents worldwide are listed in Figure 13 below:

The paper that gained prominence with 519 citations was "Blockchain technology and its relationships to sustainable supply chain management" by authors Saberi, Kouhizadeh, and Sarkis published in the year 2018. The paper examines blockchain technology and smart contracts with potential application to supply chain management.

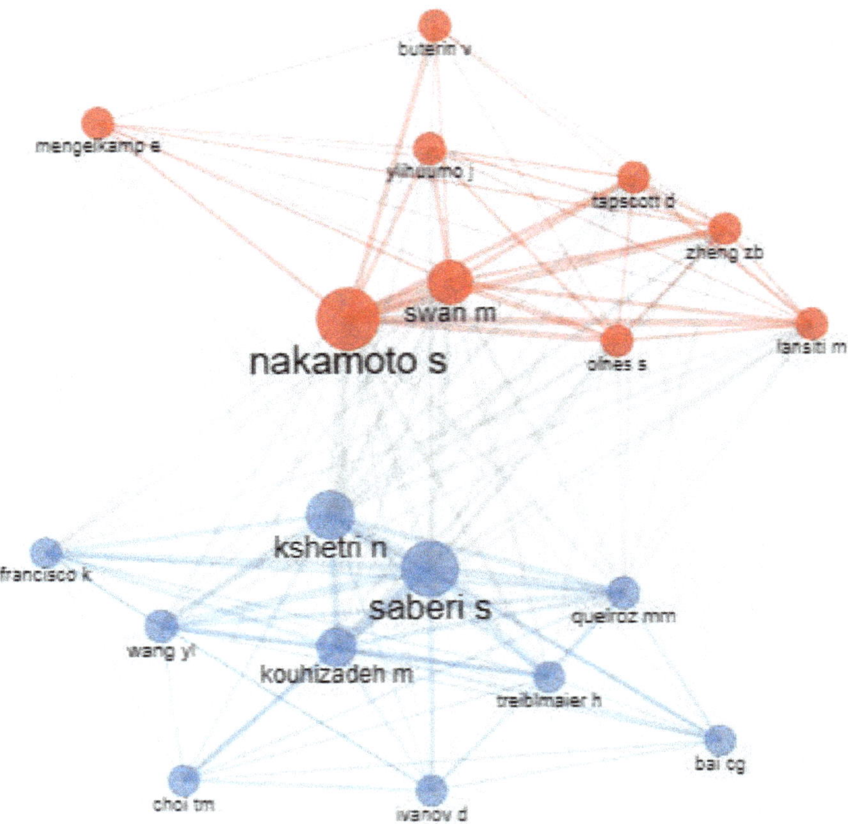

Figure 9. 20-node collaborative network of authors. Source: The authors (2021).

In this context, the twenty publications that were highlighted by the number of citations are listed in Table 4.

After the systematic analysis of the twenty most globally cited publications, as shown in Table 4, all titles and abstracts were read with the intent of answering the research question: How can the Blockchain technology aid in environmental sustainability? Of the 20 articles found, six answered the research question.

The first article that answers the question is "Blockchain technology and its relationships to sustainable supply chain management" by authors Saberi, Kouhizadeh, and Sarkis (2019), which explains that blockchain technology has the potential to contribute to social supply chain sustainability. According to the authors, making information stable and immutable is a way to build social supply chain sustainability. For Sarkis (2019), Blockchain technology can support environmental sustainability in four ways. Firstly, to accurately track sub-standard products and

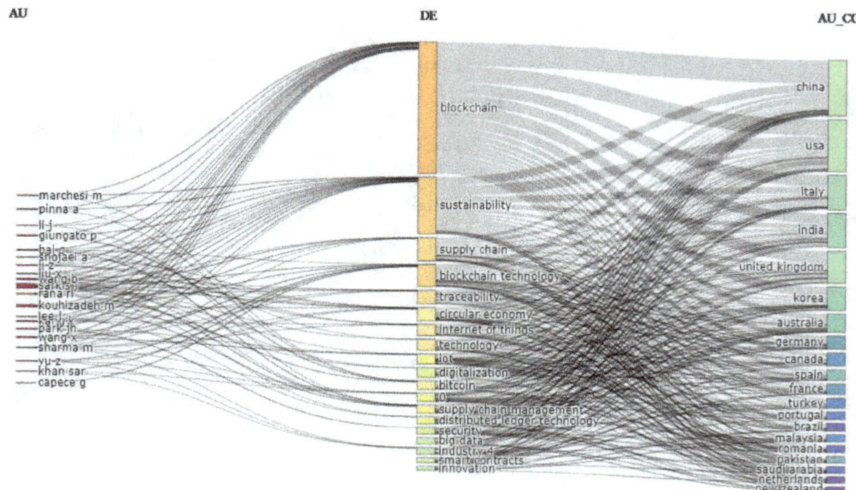

Figure 10. Three-Field Plot, with key-word of author, authors and countries. Source: The authors (2021).

Figure 11. TreeMap. Source: The authors (2021).

identify further transactions of the products, thus can help reduce the consumption of resources and greenhouse gas emissions. Secondly, Blockchain can be used to ensure that supposedly green products are indeed environmentally friendly. In this sense Blockchain technology can help reduce carbon emissions in the product journey by providing the foundation for supply chain mapping and applying low carbon product design and production. Thirdly, blockchain can improve recycling as it has been used to motivate people in northern Europe through financial

Figure 12. Tag cloud. Source: The authors (2021).

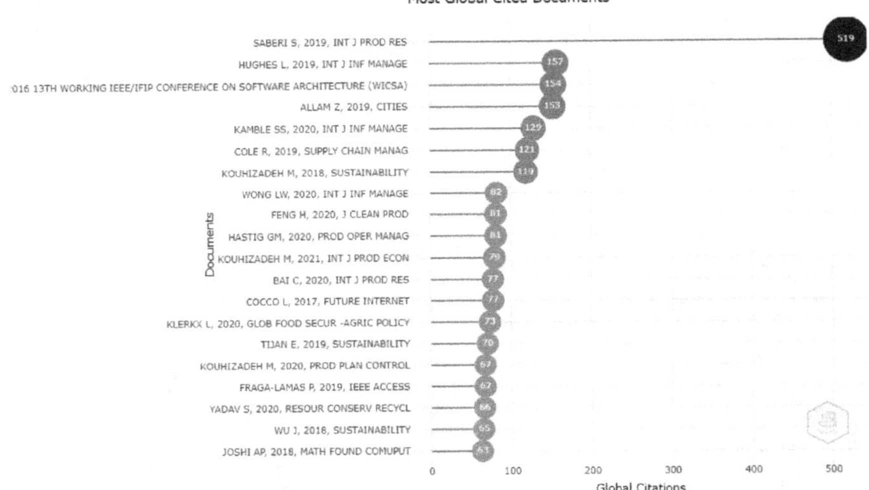

Figure 13. Most cited documents worldwide. Source: The authors (2021).

rewards in the form of cryptographic tokens in exchange for depositing recyclable materials such as plastic containers, cans or bottles. And fourthly, blockchain benefits the emissions trading process by improving the effectiveness of emissions trading schemes.

Another article is "Blockchain technology and the sustainable supply chain: Theoretically exploring adoption barriers" by Kouhizadeh et al. (2021). Consumers seek to verify whether the products they use are sustainable requesting an information portal regarding the products. This situation has put more pressure on suppliers to become more sustainable. And thus, they demand a sustainability certification system from their suppliers. As an example, there is the Business Social Compliance Initiative (BSCI) database that certifies supplier sustainability audits.

In the article entitled "Blockchain critical success factors for sustainable supply chain", the authors Yadav and Singh (2020) explain

Table 4. Most cited publications.

Year	Authors	Title	Total Number of Citations
2019	Sara Saberi, Mahtab Kouhizadeh and Joseph Sarkis	Blockchain technology and its relationships to sustainable supply chain management	519
2019	Laurie Hughes; Yogesh K. Dwivedi; Santosh K. Misra; Nripendra P. Rana; Ishnupriya Raghavan and Viswanadh Akellad	Blockchain research, practice and policy: Applications, benefits, limitations, emerging research themes and research agenda	157
2016	Xiwei Xu; Cesare Pautasso; Liming Zhu; Vincent Gramoli; Alexander Ponomarev; An Binh Tran; Shiping Chen	The Blockchain as a Software Connector	154
2019	Zaheer Allam and Zaynah A. Dhunny	On big data, artificial intelligence and smart cities	153
2020	Sachin S. Kamble; Angappa Gunasekaran and Rohit Sharma	Modeling the blockchain enabled traceability in agriculture supply chain	129
2019	R. Cole; M. Stevenson and J. Aitken	Blockchain technology: implications for operations and supply chain management	121
2018	Mahtab Kouhizadeh and Joseph Sarkis	Blockchain Practices, Potentials, and Perspectives in Greening Supply Chains	119
2020	Lai-Wan Wong; Lai-Ying Leong; Jun-Jie Hew; Garry Wei-Han Tan and Keng-Boon Ooi	Time to seize the digital evolution: Adoption of blockchain in operations and supply chain management among Malaysian SMEs	82
2020	Huanhuan Feng; Xiang Wanga; Yanqing Duan; Jian Zhang and Xiaoshuan Zhanga	Applying blockchain technology to improve agri-food traceability: A review of development methods, benefits and challenges	81
2019	Gabriella M. Hastig and Man Mohan S. Sodhi	Blockchain for Supply Chain Traceability: Business Requirements and Critical Success Factors	81
2021	Mahtab Kouhizadeh; Sara Saberi and Joseph Sarkis	Blockchain technology and the sustainable supply chain: Theoretically exploring adoption barriers	79

Table 4 contd. ...

...Table 4 contd.

Year	Authors	Title	Total Number of Citations
2020	Chunguang Baia and Joseph Sarkis	A supply chain transparency and sustainability technology appraisal model for blockchain technology	77
2017	Luisanna Cocco; Andrea Pinna and Michele Marchesi	Banking on Blockchain: Costs Savings Thanks to the Blockchain Technology	77
2020	Laurens Klerkx and David Christian Rose	Dealing with the game-changing technologies of Agriculture 4.0: How do we manage diversity and responsibility in food system transition pathways?	73
2019	Edvard Tijan; Sasa Aksentijevic; Katarina Ivanić and Mladen Jardas	Blockchain Technology Implementation in Logistics	70
2020	Mahtab Kouhizadeh; Qingyun Zhu and Joseph Sarkis	Blockchain and the circular economy: potential tensions and critical reflections from practice	67
2019	Paula Fraga-Lamas and Tiago M. Fernández-Caramés	A Review on Blockchain Technologies for an Advanced and Cyber-Resilient Automotive Industry	67
2020	Sachin Yadav and Surya Prakash Singh	Blockchain critical success factors for sustainable supply chain	66
2018	Jiani Wu and Nguyen Khoi Tran	Application of Blockchain Technology in Sustainable Energy Systems: An Overview	65
2018	Archana Prashanth Joshi; Meng Han and Yan Wang	A survey on security and privacy issues of blockchain technology	63

that blockchain technology can help sustainability through security, irreversibility, distribution, transparency in the green certification of products that provide credibility to consumers.

In the article entitled "A supply chain transparency and sustainability technology appraisal model for blockchain technology", Bai and Sarkis (2020) explain that there are three dimensions within organizations which are: technology-organization-environment (TOE). The TOE structure involving technology-organization-environment is an organization-level theory that explains that these three different elements influence decision-making concerning sustainability actions, adopted by companies.

In the article "Blockchain and the circular economy: potential tensions and critical reflections from practice", authors Mahtab et al. (2020) point out that when a record is attached to blockchain ledgers, multiple copies of the record are then shared among participants. Keeping multiple copies of transactions can further ensure the reliability and security of the information and make the information forgery-proof. If a network participant attempts to forge information and data changes, they can be traced through secure audit trails. This reliability can elevate sustainability practices, where customers can guarantee the origin, activities performed and provenance of green goods and products.

Another article is "Application of Blockchain Technology in Sustainable Energy Systems: an Overview" by authors Wu and Tran (2018). Blockchain is a type of chained data structure in which blocks of data are connected in a time sequence, and cryptography is used to ensure non-faulty modification of the distributed ledger. The definition of this technology in a broad sense is that blockchain is a new way of applying computer technology such as distributed data storage, encryption algorithms for sustainable energy management, consensus mechanisms and peer-to-peer transmission.

5. Conclusion

Blockchain is a technology that creates a point-to-point network of nodes using a worldwide network of computers, generating for each use or modification of a given document, a record that cannot be easily erased. Its use can support environmental sustainability by accurately tracking substandard products. This can decrease resource consumption as well as reduce greenhouse gas emissions. Blockchains may be used to verify that ostensibly eco-friendly items are actually such.

It was found that recycling can be improved as this technology can be used to motivate people in Northern Europe through financial rewards in the form of cryptographic tokens in exchange for depositing recyclable materials such as plastic recipients, cans or bottles. As a result, blockchain can help the emissions trading process by increasing the efficiency of existing carbon trading systems. For future research the authors propose studying blockchain as a technology for green innovation.

References

Allam, Z. and Dhunny, Z.A. (2019). On big data, artificial intelligence and smart cities. Cities (London, England), 89: 80–91. doi: 10.1016/j.cities.2019.01.032.

Antonopoulos, A.M. (2017). Mastering Bitcoin: Programming the open blockchain. "O'Reilly Media, Inc.".

Bai, C. and Sarkis, J. (2020). A supply chain transparency and sustainability technology appraisal model for blockchain technology. International Journal of Production Research, 58(7): 2142–2162. doi:10.1080/00207543.2019.1708989.

Berryhill, J., Bourgery, T. and Hanson, A. (2018). Blockchains unchained: Blockchain technology and its use in the public sector.

Bhaskar, S., Tan, J., Borges, M.L.A.M., Minssen, T., Badaruddin, H., Israeli-Korn, S. and Chesbrough, H. (2020). At the epicenter of COVID-19–the tragic failure of the global supply chain for medical supplies. Frontiers in Public Health, 8. doi: 10.3389/fpubh.2020.562882.

Bogers, M., Chesbrough, H. and Moedas, C. (2018). Open innovation: Research, practices, and policies. California Management Review, 60(2): 5–16. Disponível em: https://journals.sagepub.com/doi/full/10.1177/0008125617745086. Acesso em: 20 ago. 2021.

Bruno-Faria, M. de F. and Fonseca, M.V. de A. (2014). Cultura de Inovação: conceitos e modelos teóricos. RAC. Rio de Janeiro, 18(4): art. 1, pp. 372–396, Jul./Ago. Disponível em: https://doi.org/10.1590/1982-7849rac20141025. Acesso em: 20 jul. 2021.

Buchmann, J. (2004). Introduction to Cryptography. New York: Springer.

Carta De Clara E Francisco: Direto do Brasil para o Encontro Mundial em Assis. São Paulo, 19 nov. 2019. Disponível em: http://economiadefranciscoeclara.com.br/. Acesso em: 20 dez. 2021.

Centre For Reviews and Dissemination. (2009). Systematic Reviews: CRD's guidance for undertaking reviews in health care. York: CRD.

Chesbrough, H. (2012). Inovação aberta: como criar e lucrar com a tecnologia (LC de Q).

Cocco, L., Pinna, A. and Marchesi, M. (2017). Banking on blockchain: Costs savings thanks to the blockchain technology. Future Internet, 9(3): 25. doi:10.3390/fi9030025.

Cole, R., Stevenson, M. and Aitken, J. (2019). Blockchain technology: Implications for operations and supply chain management. Supply Chain Management: An International Journal, 24(4): 469–483. doi:10.1108/scm-09-2018-0309.

Coulouris, G. et al. (2013). Sistemas Distribuídos: Conceitos e Projeto. Bookman Editora.

Creswell, J. W. and Creswell, J.D. (2021). Projeto de pesquisa: Métodos qualitativo, quantitativo e misto.

Cusack, D.F., Kazanski, C.E., Hedgpeth, A., Chow, K., Cordeiro, A.L., Karpman, J. and Ryals, R. (2021). Reducing climate impacts of beef production: A synthesis of life cycle assessments across management systems and global regions. Global Change Biology, 27(9): 1721–1736. doi: 10.1111/gcb.15509.

De-la-Torre-Ugarte-Guanilo, M.C., Takahashi, R.F. and Bertolozzi, M.R. (2011). Revisão sistemática: noções gerais. Revista da Escola de Enfermagem USP, São Paulo, 45(5): 1260—1266, out. doi: http://dx.doi.org/10.1590/S0080-62342011000500033.

Di Pierro, M. (2017). What is the blockchain? Computing in Science & Engineering, 19(5): 92–95. doi: 10.1109/MCSE.2017.3421554.

Fedorova, E.P. and Skobleva, E.I. (2020). Application of blockchain technology in higher education. European Journal of Contemporary Education, 9(3): 552–571.

Feng, H., Wang, X., Duan, Y., Zhang, J. and Zhang, X. (2020). Applying blockchain technology to improve agri-food traceability: A review of development methods, benefits and challenges. Journal of Cleaner Production, 260: 121031. doi: 10.1016/j.jclepro.2020.121031.

Fraga-Lamas, P. and Fernandez-Carames, T.M. (2019). A review on blockchain technologies for an advanced and cyber-resilient automotive industry. IEEE Access: Practical Innovations, Open Solutions, 7(1-1): 17578–17598. doi: 10.1109/access.2019.2895302.

Global technology governance report 2021. (n.d.). World Economic Forum. Retrieved December 27, 2021, from https://www.weforum.org/reports/global-technology-governance-report-2021.

Hastig, G.M. and Sodhi, M.S. (2020). Blockchain for supply chain traceability: Business requirements and critical success factors. Production and Operations Management, 29(4): 935–954. doi:10.1111/poms.13147.

Hashimy, L., Treibelmayer, H. and Jain, G. (2021). Distributed ledger technology as a catalyst for open innovation adoption among small and medium-sized enterprises. Journal of High Technology Management Research, 32(1): 100405. doi: 10.1016/J.HITECH.2021.100405.

Herweijer, C., Combes, B., Johnson, L., McCargow, R., Bhardwaj, S., Jackson, B. and Ramchandani, P. (2018). Enabling a sustainable Fourth Industrial Revolution: How G20 countries can create the conditions for emerging technologies to benefit people and the planet (No. 2018-32). Economics Discussion Papers.

Hughes, L., Dwivedi, Y.K., Misra, S.K., Rana, N.P., Raghavan, V. and Akella, V. (2019). Blockchain research, practice and policy: Applications, benefits, limitations, emerging research themes and research agenda. International Journal of Information Management, 49: 114–129. doi: 10.1016/j.ijinfomgt.2019.02.005.

Hull, J., Gupta, A. and Kloppenburg, S. (2021). Interrogating the promises and perils of climate cryptogovernance: Blockchain discourses in international climate politics. Earth System Governance, 9: 100117. doi: 10.1016/j.esg.2021.100117.

IPCC. (2019). Climate Change and Land: An IPCC Special Report on climate change, desertification, land degradation, sustainable land management, food security, and greenhouse gas fluxes in terrestrial ecosystems (SRCCL) (Shukla, P.R., Skea, J., Calvo Buendia, E., Masson-Delmotte, V., Pörtner, H.-O., Roberts, D.C., Zhai, P., Slade, R., Connors, S., Van Diemen, R., Ferrat, M., Haughey, E., Luz, S., Neogi, S., Pathak, M., Petzold, J., Portugal Pereira, J., Vyas, P., Huntley, E., Kissick, K., Belkacemi, M. and Malley, J. (eds.)). UN Intergovernmental Panel on Climate Change.

Johnson, C., Sierra, A.R., Dettmer, J., Sidiropoulou, K., Zicmane, E., Canalis, A., Llorente, P., Paiano, P., Mengal, P. and Puzzolo, V. (2021). The bio-based industries joint undertaking as a catalyst for a green transition in Europe under the European Green Deal. EFB Bioeconomy Journal, 1: 100014. doi: 10.1016/j.bioeco.2021.100014.

Kamble, S.S., Gunasekaran, A. and Sharma, R. (2020). Modeling the blockchain enabled traceability in agriculture supply chain. International Journal of Information Management, 52: 101967. doi: 10.1016/j.ijinfomgt.2019.05.023.

Klerkx, L. and Rose, D. (2020). Dealing with the game-changing technologies of Agriculture 4.0: How do we manage diversity and responsibility in food system transition pathways? Global Food Security, 24: 100347. doi: 10.1016/j.gfs.2019.100347.

Kouhizadeh, M. and Sarkis, J. (2018). Blockchain practices, potentials, and perspectives in greening supply chains. Sustainability, 10(10): 3652. doi: 10.3390/su10103652.

Kouhizadeh, M., Zhu, Q. and Sarkis, J. (2020). Blockchain and the circular economy: Potential tensions and critical reflections from practice. Production Planning & Control, 31(11-12): 950–966. doi: 10.1080/09537287.2019.1695925.

Kouhizadeh, M., Saberi, S. and Sarkis, J. (2020). Blockchain technology and the sustainable supply chain: Theoretically exploring adoption barriers, International Journal of Production Economics. doi: https://doi.org/10.1016/j.ijpe.2020.107831.

Kouhizadeh, M., Saberi, S. and Sarkis, J. (2021). Blockchain technology and the sustainable supply chain: Theoretically exploring adoption barriers. International Journal of Production Economics, 231: 107831. doi: 10.1016/j.ijpe.2020.107831.

Lewis, A. (2018). The basics of bitcoins and blockchains: an introduction to cryptocurrencies and the technology that powers them. Mango Media Inc.

Li, G. and Fang, C. (2021). Promoting information-resource sharing within the enterprise: A perspective of blockchain consensus perception. Journal of Open Innovation: Technology, Market, and Complexity, 7(3): 177–195. doi: 10.3390/joitmc7030177.

Linnenluecke, M.K., Marrone, M. and Singh, A.K. (2020). Sixty years of Accounting & Finance: A bibliometric analysis of major research themes and contributions. Accounting & Finance, 60(4): 3217–3251. doi: 10.1111/acfi.12714.

Machado, A. de bem and Richter, M.F. (2020). Sustentabilidade Em Tempos De Pandemia (COVID-19): (Covid-19). RECIMA21 - Revista Científica Multidisciplinar - ISSN 2675-6218, 1(2), 264–279. https://doi.org/10.47820/recima21.v1i2.25.

Mancebo, F. and Sachs, I. (eds.). (2015). Transitions to Sustainability. 162 pp. doi: 10.1007/978-94-017-9532-6.

Meyer, C., Gerlitz, L. and Henesey, L. (2021). Cross-border capacity-building for port ecosystems in small and medium-sized baltic ports. TalTech Journal of European Studies, 11(1): 113–32. doi: 10.2478/bjes-2021-0008.

Morakanyane, R., Grace, A.A. and O'Reilly, P. (2017). Conceptualizing digital transformation in business organizations: A systematic review of literature. Procedimentos BLED 2017. 21. Disponível em: https://aisel.aisnet.org/bled2017/21. Acesso em: 31 jul 2021.

Mosteanu, N.R. and Faccia, A. (2021). Fintech frontiers in quantum computing, fractals, and blockchain distributed ledger: Paradigm shifts and open innovation. Journal of Open Innovation: Technology, Market, and Complexity, 7(1): 1–19. doi: 10.3390/joitmc7010019.

Nakamoto, S. (2008). Bitcoin: A peer-to-peer electronic cash system. Decentralized Business Review, 21260.

Nobre, I. and Nobre, C. (2020). Amazon 4.0. Futuribles, 1: 95–108.

Ozdemir, A.I., Erol, I., Ar, I.M., Peker, I., Asgary, A., Medeni, T.D. and Medeni, I.T. (2020). The role of blockchain in reducing the impact of barriers to humanitarian supply chain management. The International Journal of Logistics Management, 32(2): 454–478. doi: 10.1108/IJLM-01-2020-0058.

Pazaitis, A. (2020). Breaking the Chains of open innovation: Post-blockchain and the case of sensorica. Information, 11(2): 104–121. doi: 10.3390/info11020104.

Peters, B.G. (2017). What is so wicked about wicked problems? A conceptual analysis and a research program. Policy and Society, 36(3): 385–396. doi: 10.1080/14494035.2017.1361633.

Prashanth Joshi, A., Marietta, H.M. and Wang, Y. (2018). A survey on security and privacy issues of blockchain technology. Mathematical Foundations of Computing, 1(2): 121–147. doi:10.3934/mfc.2018007.

Purvis, B., Mao, Y. and Robinson, D. (2019). Three pillars of sustainability: In search of conceptual origins. Sustainability Science, 14(3): 681–695. doi: 10.1007/s11625-018-0627-5.

Quaresma, P. and Lopes, E. (2021). Criptografia. Departamento de Matemática, Universidade de Coimbra. Disponível em: <https://www.mat.uc.pt/~pedro/lectivos/CodigosCriptografia1213/artigo-gazeta08.pdf>. Acesso em: 02 de nov. de 2021.

Rittel, H.W.J. and Webber, M.M. (1973). Dilemmas in a general theory of planning. Policy Sciences, 4(2): 155–169.

Rotabi, S. and Ali, O. (2022). Applications of blockchain technology for a circular economy with focus on Singapore. In Blockchain Technologies for Sustainability (pp. 151–178). Springer, Singapore.

Saberi, S., Kouhizadeh, M., Sarkis, J. and Shen, L. (2019). Blockchain technology and its relationships to sustainable supply chain management. International Journal of Production Research, 57(7): 2117–2135. doi: 10.1080/00207543.2018.1533261.

Schletz, M., Franke, L.A. and Salomo, S. (2020). Blockchain application for the Paris agreement carbon market mechanism—A decision framework and architecture. Sustainability, 12(12): 5069. doi: 10.3390/su12125069.

Schwab, K. and Davis, N. (2018). Shaping the future of the fourth industrial revolution. Currency.

Seuring, S., Sarkis, J., Müller, M. and Rao, P. (2008). Sustainability and supply chain management–An introduction to the special issue. Journal of Cleaner Production, 16(15): 1545–1551. doi: 10.1016/j.jclepro.2008.02.002.

Swan, M. (2015). Blockchain: Blueprint for a New Economy. O'Reilly Media, Inc.

Tidd, J. and Bessant, J. (2015). Gestão da Inovação. Tradução de Félix Nonnenmacher. 5. ed. Porto Alegre: Bookman.

Tijan, E., Aksentijević, S., Ivanić, K. and Jardas, M. (2019). Blockchain technology implementation in logistics. Sustainability, 11(4): 1185–1197. doi: 10.3390/su11041185.

Techopedia. (2011, July 25). Cryptography. Techopedia.Com; Techopedia. https://www.techopedia.com/definition/1770/cryptography.

Thangavel, P. and Sridevi, G. (2016). Environmental Sustainability. Springer, India, Private.

Thormann, L., Neuling, U. and Kaltschmitt, M. (2021). Opportunities and challenges of the european green deal for the chemical industry: An approach measuring innovations in bioeconomy. Resources, 10(9): 91. doi: 10.3390/resources10090091.

Universidade De York. Centro de Avaliações e Divulgação. Disponível em: https://www.york.ac.uk/crd/about/. Acesso em: 17 out. 2021.

Walch, A. (2016). The path of the blockchain lexicon (and the law). Boston University Review of Banking & Financial Law, 36: 713–725.

Whittemore, R. and Knafl, K. (2005). The integrative review: Updated methodology. Journal of Advanced Nursing, 52(5): 546–553.

Wong, L.-W., Leong, L.-Y., Hew, J.-J., Tan, G.W.-H. and Ooi, K.-B. (2020). Time to seize the digital evolution: Adoption of blockchain in operations and supply chain management among Malaysian SMEs. International Journal of Information Management, 52: 101997. doi: 10.1016/j.ijinfomgt.2019.08.005.

Wu, J. and Tran, N. (2018). Application of blockchain technology in sustainable energy systems: An overview. Sustainability, 10(9): 3067. doi: 10.3390/su10093067.

Xu, X., Pautasso, C., Zhu, L., Gramoli, V., Ponomarev, A., Tran, A.B. and Chen, S. (2016). The blockchain as a software connector. 2016 13th Working IEEE/IFIP Conference on Software Architecture (WICSA). IEEE.

Yadav, S. and Singh, S.P. (2020). Blockchain critical success factors for sustainable supply chain. Resources, Conservation, and Recycling, 152: 104505. doi: 10.1016/j.resconrec.2019.104505.

CHAPTER 9

Cryptocurrencies—Advantages and Risks of Digital Money[1]

Gabriel Osório de Barros

1. Introduction

Historically, humanity performed exchange of goods but the introduction of money has further facilitated the process of exchange. The way money is transacted has evolved through different forms of transaction - currency, banking transactions, credit cards, etc. The increasing digitalization of society led to the emergence of electronic money. There is an increasing interest in this type of digital money, particularly based on cryptography - called cryptocurrencies. The creation of bitcoin (Nakamoto, 2008) was an important driver behind blockchain technology, which has increased applications in diverse fields.

Cryptocurrencies are a type of digital or virtual currency that uses cryptography to facilitate secure and decentralized financial transactions. They are based on decentralized systems that use blockchain technology

GEE - Office for Strategy and Studies, Ministry of Economy and Maritime Affairs, Portugal.
Email: gabriel.barros@gee.gov.pt

[1] This article is the sole responsibility of the author and does not necessarily reflect the positions of the Office for Strategy and Studies or the Portuguese Ministry of Economy and Maritime Affairs. A previous version of this article was initially published as an Economic Theme by the Office for Strategy and Studies in a series of articles neither paper nor working paper, and publication as a chapter has the agreement of the office's management.

to record and verify transactions. Cryptocurrencies are different from traditional fiat currencies, which are issued and backed by central governments.

Unlike traditional fiat money, which is issued and controlled by central banks, cryptocurrencies are not subject to the same level of regulation and oversight. They offer several potential advantages, such as low transaction fees, fast and global transaction processing, and the ability to facilitate anonymous transactions. However, they also come with risks, including high price volatility, lack of legal protection for consumers, and the potential for fraud or hacking. Despite these challenges, cryptocurrencies have gained a significant interest and have demonstrated their potential as a disruptive force in the financial industry.

This paper explores the history, advantages, and risk of cryptocurrencies, as well as the current state and potential future developments in the field. The goal of the paper is to provide a balanced analysis of the advantages and risks of using cryptocurrencies as a form of digital money.

2. History of Cryptocurrencies

Cryptocurrencies, from the outset, have been associated with the liberalist concern which believes State intervention in financial transactions should be as little as possible. The emergence in the 1990s of Cypherpunks[2] and Crypto-anarchists[3] movements that considered it essential to guarantee privacy in transactions by using cryptography and having less participation of the State, had great impact in the creation of cryptocurrencies.

Concerns about how to deal with the movement of money through cryptography led to the initial studies, of which the works of Chaum (1985, 1988) stand out. The concern for privacy[4] is central in these studies and Chaum focuses on the creation of "unconditionally untraceable electronic money" (Chaum et al., 1988).

However, unlike physical money, this type of solution presents difficulties because it is necessary to avoid creating multiple copies of electronic money (in physical currencies it is much more difficult to create exact copies).

[2] https://www.activism.net/cypherpunk/manifesto.html.

[3] https://www.activism.net/cypherpunk/crypto-anarchy.html.

[4] "The use of credit cards today is an act of faith on the p a t of all concerned. Each party is vulnerable to fraud by the others, and the cardholder in particular has no protection against surveillance." Chaum et al., 1988.

Dwork and Naor (1992) presented a computational technique to control access to shared resources and aiming to combat junk emails. In 1997, Back (2002) proposed a similar function which he called HashCash.[5]

In 1998, Wei Dai wrote a proposal on B-Money,[6] although he did not continue with his study. In 2008, Satoshi Nakamoto sent an email to Wei Dai in which he referred to his interest in the study published on the B-Money page and said he would release a paper that turned the idea into a "complete working system".[7] Two months later, Nakamoto (2008)[8] published the white paper on Bitcoin, with the design and motivation for the creation of a currency that is not controlled by any entity, based on cryptography.

The white paper thus laid the foundations for an open, accessible, non-centralized currency, independent of financial intermediaries (the network acts as a substitute for banks), and where user privacy is guaranteed. The same document also laid the foundations of blockchain technology.

With Bitcoin, anyone with a computer and internet connection can join the network. Minting and distribution of bitcoins is carried out through mining,[9] in a process that aims to be decentralized.

In 2009, the Bitcoin source code was launched, and the bitcoin network started with the mining of the first bitcoins.

Bitcoin is considered the first successful cryptocurrency, being the largest, most used, and best known. Subsequently, many other cryptocurrencies were created (known as altcoins - alternative coins) that also use the blockchain technology, standing out in terms of market cap, with Bitcoin (Nakamoto, 2008), Ethereum (Buterin, 2014) and Cardano (Hoskinson, 2017) standing out.

On January 15, 2023, the top five cryptocurrencies had a market value around 720 billion USD (Table 1). Notwithstanding, this amount is

---

[5] "Hashcash used the cryptographic hash function SHA-1 (Secure Hash Algorithm 1) to create a stamp that would help in verifying to the recipient that the email was not spam" (Lielacher, 2018).

[6] http://www.weidai.com/bmoney.txt.

[7] https://www.gwern.net/docs/bitcoin/2008-nakamoto.

[8] To this day, Nakamoto's identity remains unknown.

[9] "The records are grouped and stored in blocks. Each block contains a timestamp and a link to a previous block so that the blocks are chained together, thus the name blockchain. The blocks are mined in sequence, and once recorded, the data cannot be altered retroactively. A complete record of transactions can be found on the main chain. Each block on the chain is linked to the previous one and can be traced all the way back to the very first block, which is called the genesis block. However, there are also blocks that are not part of the main chain, called detached or orphaned blocks. They can occur when more than one miner produces blocks at similar times, or they can be caused by attackers' attempt to reverse transactions. When separate blocks are validated concurrently, the algorithm will help maintain the main chain by selecting the block with the highest value." (Lee, 2018).

Table 1. Top 5 cryptocurrencies (by USD market value) on January 15, 2023.

#	Name	Symbol	Market Cap (USD)
1.	Bitcoin	BTC	385,086,196,581
2.	Ethereum	ETH	180,902,207,502
3.	Tether	USDT	66,210,931,564
4.	Binance Coin	BNB	46,511,492,058
5.	USD Coin	USDC	42,209,620,127

Source: https://www.cryptocurrencychart.com/.

Table 2. Top 5 cryptocurrencies (by USD market value) on November 3, 2021.

#	Name	Symbol	Market Cap (USD)
1.	Bitcoin	BTC	1,237,302,183,654
2.	Ethereum	ETH	557,772,179,857
3.	Tether	USDT	106,029,876,159
4.	Binance Coin	BNB	73,518,076,260
5.	USD Coin	USDC	34,417,492,232

Source: https://www.cryptocurrencychart.com/.

significantly lower than the market value of the top five cryptocurrencies on November 3, 2021, which was about 2,009 billion USD (Table 2).

Figure 1 shows that the market cap of the top 5 cryptocurrencies changed significantly over time.

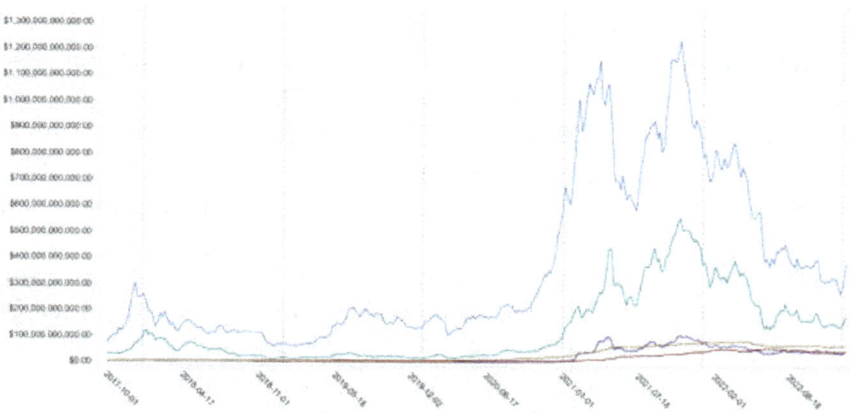

Figure 1. Evolution of the market cap of the top 5 cryptocurrencies (in USD) from October 1, 2017, until January 15, 2023.

Source: https://www.cryptocurrencychart.com/ .

With the proliferation of Bitcoin, other platforms that allow the commercialization of bitcoins without the need to join the Bitcoin network have emerged, namely Coinbase, Bitpay, blockchain.info, as well as other cryptocurrencies, some of which have not been successful.

3. "Old Money" and the Cryptocurrencies

Money is used in exchanges, for example to pay for goods and services, to purchase other currencies or for financial transactions, and is generally materialized in the form of banknotes and coins. More recently, digital money, namely cryptocurrencies, emerged with specific characteristics.

Usually, the following characteristics are attributed to money:

- It is a common denomination of value, allowing the comparison of the value of different goods and services;
- It is a means of payment;
- It is divisible into subunits;
- It is portable, allowing easy transportation;
- It is widely accepted;
- It has a relatively stable value;
- It is a reserve of value, allowing savings;
- It is controlled by the central banks of the countries;
- It is hardly reproducible.

Some of these characteristics are common to both types of money but other are distinctive. Considering cryptocurrencies:

- As a unit of measure, they allow, like traditional money, the comparison of the value of different goods and services, although the volatility may prevent comparison at different times;
- Like traditional physical money, it is a means of payment;
- Since they exist as global currencies, their value is the same anywhere in the world and no exchange rate is required (except in exchange for other currencies), facilitating international trade;
- They are also divisible, like traditional currencies, into sub-units, but, unlike traditional currencies, they allow dividing up to 8 decimal places (the smallest portion being called Satoshi);
- Their portability is even greater than the physical money, being necessary only to have the keys for access;
- However, their acceptance is not generalized and there is still a limited set of entities that accept it, although the number of people

and companies that accept them as means of payment is growing (as referred later);

- As for value stability, cryptocurrencies have shown great instability, generally justified by the fact that it is still an emerging technology at an early stage, and does not currently have stability similar to that of traditional money;
- Exactly because of the volatility issue and because cryptocurrencies have no intrinsic value, it is questionable if they can be considered a reserve of value;[10]
- In the case of cryptocurrencies, for the very reasons that justified their creation, there is no control by central banks and no supervision;
- Unlike bank notes, that are hard to manufacture as counterfeit currency, cryptocurrencies have keys that may be copied if left unprotected.

Overall, we can highlight the following differences between traditional fiat money and cryptocurrencies:

- Cryptocurrencies can be mined or validated through the process of blockchain, while traditional fiat money is created and distributed through a central banking system;
- Fiat money is issued and controlled by central banks, while cryptocurrencies are decentralized and not controlled by any central authority;
- Cryptocurrencies offer a higher level of anonymity for transactions, while traditional fiat money can be traced through financial institutions;
- Cryptocurrencies are not subject to regulation and oversight, unlike traditional fiat money;
- Cryptocurrencies use cryptography to secure transactions, while traditional fiat money relies on physical or digital security measures;
- Traditional fiat money is widely accepted and has a relatively stable value, while cryptocurrencies have a volatile value and are not yet widely accepted.

Throughout the present work, these issues will be analysed in more depth.

[10] This is also true for "normal" money, especially for currencies of countries that are exposed to hyperinflation.

4. Bitcoin Revolution

As previously referred, the cryptocurrencies are a form of digital payment, with many characteristics like traditional money (notes and coins). Cryptocurrencies use encryption in their operation and are increasingly used in the economy.

Among cryptocurrencies, it is necessary to emphasize the role of the Bitcoins as the first and most used cryptocurrency.

Since its inception on January 3, 2009, the number of Bitcoins in circulation has risen substantially, nearing 19.3 million Bitcoins, as illustrated in Figure 2. It's worth noting that the Bitcoin protocol limits the total supply to 21.0 million.

The market price of Bitcoin has fluctuated overtime and has reached record values in November 2021, having decreased significantly since then, as seen in Figure 3.

Although Bitcoin distinguishes itself from banks by guaranteeing the confidentiality of users (although not completely), it intends to have bank-like functions, ensuring that only the owner of an account can access it, keeping a record of transactions and allowing users to manage their accounts.

(total number of mined bitcoin circulating on the network)

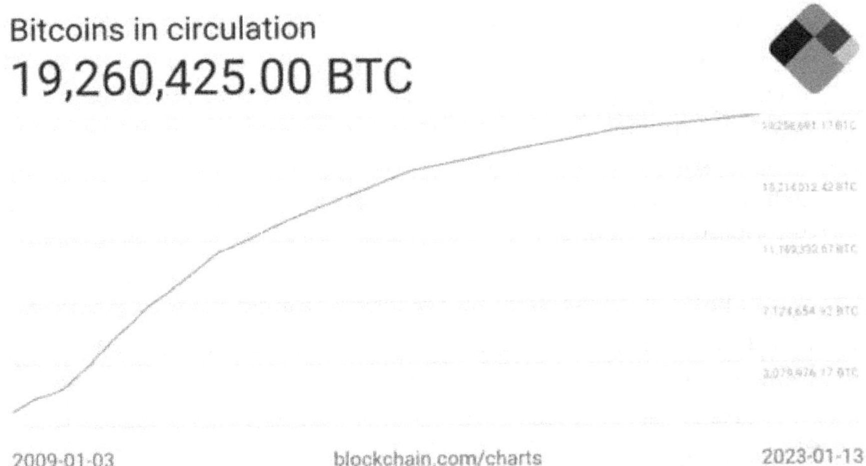

Figure 2. Bitcoins in circulation.[11]
Source: https://www.blockchain.com/en/charts/total-bitcoins.

[11] The total number of bitcoins that have already been mined, i.e., the current supply of bitcoins on the network (https://www.blockchain.com/charts/total-bitcoins).

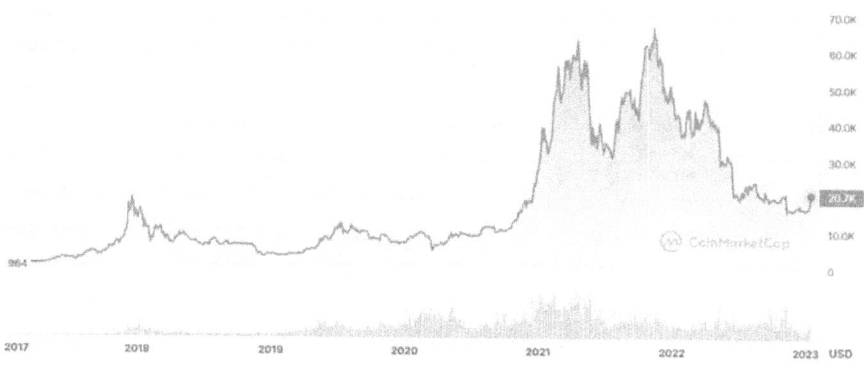

Figure 3. Bitcoin market price (USD).
Source: https://coinmarketcap.com/currencies/bitcoin/.

The reliability associated with Bitcoin is connected to how users' identities are guaranteed, each of which manages its public and private keys to send and receive money. On the one hand, public keys are used to transfer funds and private keys are randomly generated allowing to prove ownership of the public keys. On the other hand, to conduct a transaction it is necessary to ensure that funds are available and there can be no other transaction to move the same funds using an Unspent Transaction Output (UTXO). This ensures that the already-expended outputs will not be re-expended and that acts as input to a new transaction. Bitcoin stores the transactions in blocks built from its previous blocks and in this way forms the blockchain. The detailed process of a blockchain transaction can be visualized in Figure 4.

5. Blockchain

Blockchain is a data structure that records transactions of cryptocurrencies in so-called blocks, where new transactions are recorded as new blocks that join existing ones (each block is connected to a previous block) and each time a transaction is logged, it is very difficult to reverse.

As mentioned earlier, blockchain came to revolutionize cryptocurrencies and is currently the basis of existing electronic currencies.

The trust given to the blockchain results from its transparency (the information is public) and its independency (it does not depend on any entity). The operation of blockchain seeks to prevent the counterfeiting of coins, avoiding the reversal of transactions.

Blockchain and the other main cryptocurrencies are based on three essential principles - Decentralization, Pseudonymity and Immutability – from which results the reliability that is attributed to them.

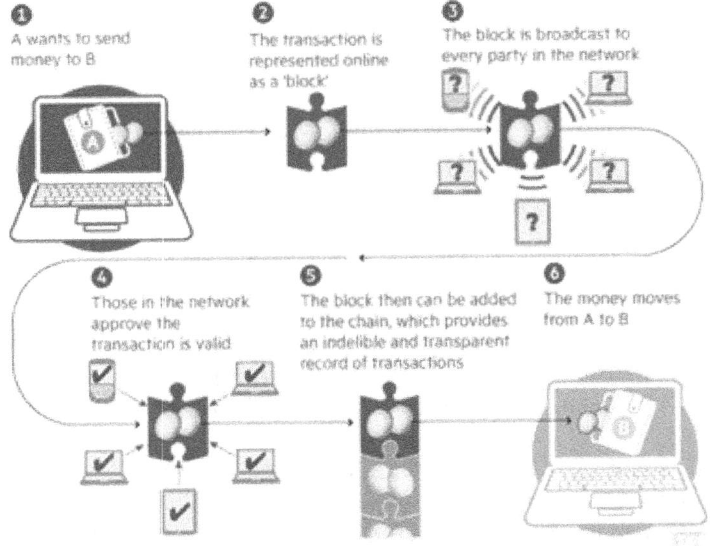

Figure 4. Blockchain transaction.

Source: Financial Times. (https://assets.weforum.org/wp-content/uploads/2015/12/151103-blockchain-bitcoin-technology-banking-fintech-FT.png).

In this chapter, we will focus on the key principles of the technology, mentioning the use of cryptographic hashes and proof-of-work for consensus and security. We will also analyse the transparency and public nature of the blockchain, as well as the risk of identity theft and the importance of maintaining integrity of information.

5.1 Decentralization

For a transaction to take place, it is necessary to verify consensus through proof-of-work that, in the absence of a central authority, limits the voting power of each user, seeking to restrict the voting capacity of malicious entities.

The process occurs through peer validation in which the user who intends to carry out the transaction sends the request to the entire network and in which everyone can validate the transaction and in which only after receiving a majority of votes the transaction is accepted.

In this way, the decision occurs in a decentralized way. This process allows identifying discrepancies between different versions of the database, facilitating consistency among all members of the network.

Decentralization helps to ensure the integrity and security of blockchain by distributing the ledger among multiple nodes. This makes it difficult for a single entity to manipulate data or to bring the network down, as it would require the coordination and cooperation of a significant number of nodes. Additionally, the use of proof-of-work or other consensus mechanisms helps to ensure that decisions about the state of the network are made in a fair and impartial manner.

5.2 *Anonymity/Pseudonymity*

As mentioned earlier, the handling of identity of the users is one of the specific characteristics of the cryptocurrencies. In fact, one of the benefits of using it is that it allows for anonymity which is not possible with traditional banks. Traditionally, banking information is associated with the identity of the customer. In cryptocurrencies, it is particularly difficult to establish this connection. For this reason, since it is not necessary to provide information about the buyer, the risk of identity theft in online payments is avoided.

As with traditional money, the use of cryptocurrencies always requires some form of authentication to prevent others from transacting with money that does not belong to them by associating the money with its owner. On the other hand, it is also necessary to keep track of transactions so that undue access can be identified and prevent malicious activity in the future. Finally, it is essential to maintain the integrity of the information, preventing the manipulation of the transactions carried out through false authentication, copying of the identity or distortion of its contents.

In this way, as in a signature, the identity of each person is guaranteed through the existence of a public key to receive crypto-coins and a private key that is extremely difficult to obtain by any other person and which must be kept confidential. In this way, the private key gives access to the public key to prove the property. From the private key the public key is automatically generated; this guarantees the unique identity of the user, and the probability of assigning the same identity to two different users is extremely reduced. This combination guarantees the identity of the user and allows the owner to transact the crypto-coins, replacing the existence of a central authority that traditionally would guarantee the unique identity of each user.

Although it is difficult to identify the user, some authors argue that it is not impossible to associate users' virtual identity with their real identity (Narayanan et al., 2016) and that anyone who knows the identity of a user in any transaction can get information about other transactions made with the same pseudonym/address (Böhme, 2015) - hence the name pseudonymity.

Blockchain networks are specifically designed to provide a higher level of anonymity and privacy to users. These technologies allow users to prove the validity of a transaction without revealing any additional information, making it even more difficult to trace the transaction back to a specific individual. However, it is important to note that these technologies are not foolproof and that it is still possible, though challenging, for a motivated actor to deanonymize users on these networks.

5.3 Immutability

As previously referred, decentralization seeks to replace a role that is usually played by a central entity. This decentralization presupposes the dissemination of a copy of all transactions to all participants in the network.

This decentralization allows the immutability of information since it is very difficult to subsequently reverse a decision because this would imply changing the information of all the users.

Immutability is achieved using cryptographic hashes. Each block in the blockchain contains a cryptographic hash of the previous block, as well as a timestamp and transaction data. This hash is a unique, fixed-size string of characters that is generated using a mathematical function that takes an input and produces an output of a fixed size. The output of this function, the hash, is unique to the input, and any change to the input will produce a completely different output.

The use of cryptographic hashes and the linked structure of the blocks in the blockchain create a chain of custody for the data that is recorded on the network. This chain of custody, or history, of the data is transparent and public, making it easy to verify the authenticity and integrity of the data at any point in time. This feature of the blockchain may make it a useful tool for a wide range of applications, including supply chain management, voting systems, and asset tracking, among others.

6. Key Approaches to Applying Blockchain in Open Innovation Processes

While blockchain has emerged as the basic technology solution for the cryptocurrencies, the technology has been adapted to enable new applications in areas such as energy, industry, and finance. For a broader spectrum of its potential applications, refer to Figure 5.

Blockchain may allow greater security and decentralization in the storage and transfer of information, seeking to ensure privacy.

Various possibilities have been identified for the application of blockchain by using decentralized information to make markets more democratic, avoiding over-concentration of market power in a small group of people by shifting control of information from business to consumers.

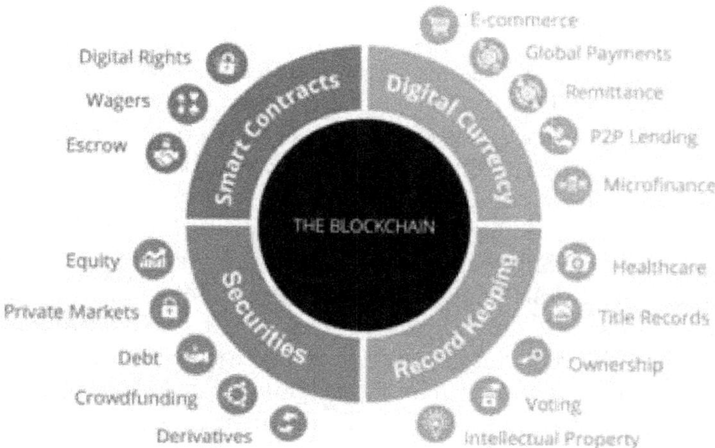

Figure 5. Blockchain's potential application and disruption.
Source: Equinix (https://blog.equinix.com/blog/2017/10/05/blockchain-a-new-type-of-internet/).

Although it's use is recent and may take years to have full effect, companies are already working on the adoption of blockchain technology,[12] seeking to simplify and strengthen their value chains.

The blockchain has the potential to enable new applications in a variety of industries. According to Bansal (2017), the application of the blockchain could allow considerable advances in sectors such as Financial Services, Automotive, Voting, Healthcare, Decentralized Notary or Smart Contracts, increasing security and decentralization in the storage and transfer of information and making markets more democratic by shifting control of information from businesses to consumers.

In fact, blockchain might become a means to avoid fraud by proving the identity of the retailer and the consumer, monitoring the movement of products, and creating verifiable registrations.

While the adoption of blockchain technology is still in its early stages and it may take years for it to reach its full potential, many companies and organizations are already working on the development and implementation of blockchain-based solutions.

[12] List of top 50 companies exploring blockchain technology available at https://www.forbes.com/sites/michaeldelcastillo/2018/07/03/big-blockchain-the-50-largest-public-companies-exploring-blockchain/#79c2bbe12b5b.

7. Practical Use

The use of cryptocurrencies makes it much easier to buy and sell goods and services, allowing to quickly transfer money.

Although in some cases cryptocurrencies were created for specific purposes, the main cryptocurrencies (Bitcoin and Ethereum) were created with the aim of having a more general use.

In terms of acceptability, there are currently a limited number of entities that allow its use. Still, the number of possibilities for cryptocurrency applications is increasing in retail businesses (e.g., sale of food or computer products) and in large consultancy companies (e.g., the "Big 4").

Although the reduced acceptance of cryptocurrencies, Jonker (2017) believes that there is substantial interest among retailers in accepting this type of currency in the future, indicating that their acceptance may increase rapidly. According to the same author, the main factors influencing the adoption of cryptocurrencies are the demand for this form of payment by consumers (one reason for non-adoption has been the low consumer demand), the added value resulting from the use (which can be influenced by the positive experience of other retailers) and non-financial barriers (retailers consider the compatibility between the use of virtual currencies and current business processes to be extremely important).

On the contrary, expectably, recent warnings from European institutions, such as the European Bank Authority and European Central Bank, and nationals such as the Bank of Portugal, cast doubt on the credibility of the cryptocurrencies and could hinder the increase in goods and services that can be paid through these coins (casting doubts on traders as to whether they should accept this form of payment).

Bitcoin is the currency that is most easily usable today, in a range of areas such as Crowd funding, Bookstores, Newspapers, Music, Movies, TV shows, Gaming, Commodities, E-commerce, Electronics, Second-hand goods, Food and Beverages, Real estate, Financial accounts, Travel and Hotels, Pharmacy, Dating, Adult content, Cloud data file services, SaaS products or Technology and Software.

Here are some ways to search for online and physical stores where cryptocurrencies may be used (non-exhaustive list).

7.1 Buying Online

One of the possibilities for using cryptocurrencies is through the purchase of goods and services at online stores. Here are some websites that enable this search.

CoinPayments

The website Coin Payments (https://www.coinpayments.net) is a directory of online operators that accept transactions in multiple cryptocurrencies.

Spendabit

Spendabit (https://spendabit.co/) is a search engine that allows searching for products sold in merchants and online platforms that can be purchased with Bitcoins.

Purse

The site Purse.io (https://purse.io) connects buyers and holders of gift cards, allowing purchases at Amazon with bitcoins (and with discount) and allowing the balance to be used to purchase goods from third parties receiving bitcoins in exchange.

7.2 *Buying at a Local Merchant*

More difficult is finding local merchants that accept bitcoins. There are, however, several sites on which searches can be performed.

Coinmap

The Coinmap website (https://coinmap.org/) allows you to identify on the map establishments that accept bitcoins as payment method.

At the geographical level, Portugal still has few options in terms of physical stores, mostly concentrated in Lisboa, Porto and Faro.

SpendBitcoins

The SpendBitcoins site (http://spendbitcoins.com/) allows you to identify physical stores (but also online) where there are merchants that accept bitcoins.

CoinATMRadar

Although it is not a site for buying goods and services, Coin ATM Radar (https://coinatmradar.com/) allows you to identify "Coin ATM" in which you can buy and sell various crypto coins (not just bitcoins) in exchange for conventional money.

7.3 *Investing and Trading*

Cryptocurrencies have gained popularity in recent years as investment and trading assets.

There are several ways to invest in or trade cryptocurrencies. One option is to buy and hold them for the long term, in the hope that they will increase in value over time. This is known as "holding." Another option is to engage in short-term trading, buying and selling cryptocurrencies quickly in an attempt to capitalize on price movements.

To invest in or trade cryptocurrencies, you will need to set up a digital wallet to store your coins and an account with a cryptocurrency exchange, such as Coinbase or Binance, where you can buy and sell cryptocurrencies.

It is important to thoroughly research and compare different exchanges and wallets before deciding, as they can vary significantly in terms of fees, security, and available coins.

Regardless of the motivations, it is important for investors and traders to understand the unique characteristics and risks of cryptocurrencies before deciding to invest or trade using them.

It is also important to be aware of the risks involved with investing in or trading with cryptocurrencies. Prices can be highly volatile and are influenced by a wide range of factors, including market demand, regulatory changes, and security issues. Cryptocurrencies are not backed by any physical assets or governments, and there is no guarantee that anyone will be able to sell them or recoup the investment. As with any investment, it is important to carefully consider risk tolerance and financial goals before deciding to invest in cryptocurrencies.

8. Volatility

One of the problems associated with cryptocurrencies is their volatility. In fact, they attracted many buyers not only because of the possibility of escaping the traditional means of currency controlled by the financial system but also because they anticipated high gains. However, cryptocurrencies have been very volatile. This becomes evident when compared with other indexes or reference prices such as S&P 500 and gold as demonstrated in Figures 6–9.

Historical volatility measures the variation over a given period. In the present study were considered the closing values of two of the main cryptocurrencies, Bitcoin and Ethereum, which were compared with the evolution of the S&P500 index and the price of gold in the period between August 7, 2015, and January 15, 2023.

The information in previous graphs shows that there has been a great instability in the cryptocurrencies, registering what appears to be "bubble-crashes" in Bitcoin and Ethereum in the beginning of 2018, around the second quarter of 2021 and during most of the year 2022.

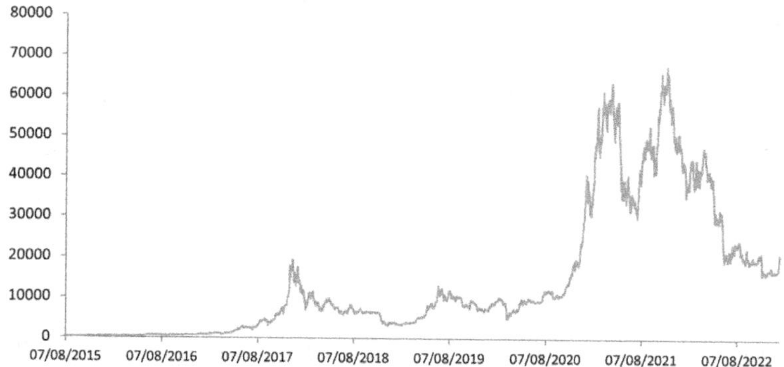

Figure 6. Evolution of the price of the Bitcoin cryptocurrency (Adjusted Close, USD/BTC).
Source: https://finance.yahoo.com/quote/BTC-USD.

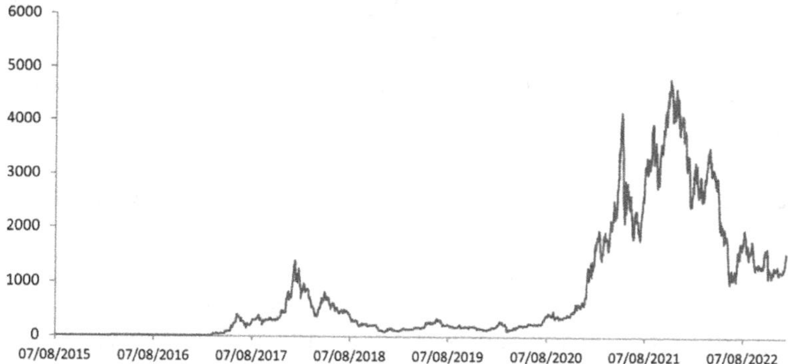

Figure 7. Evolution of the price of the Ethereum cryptocurrency (Adjusted Close, USD/ETH).
Source: https://finance.yahoo.com/quote/ETH-USD.

Figure 8. Evolution of the S&P 500 (Close, Index).
Source: https://www.spindices.com/indices/equity/sp-500.

Figure 9. Evolution of the price of gold (Closing Price, USD/oz).
Source: https://markets.businessinsider.com.

In this study, we have calculated volatility with the following calculations:

1. Calculation of ln(Pi/Pi-1), where ln is the natural logarithm, Pi is the daily reference price on day i and Pi-1 is the price relative to the previous day;
2. Calculation of the standard deviation of the series calculated in point 1 for the n days considered;
3. Annualization of the value of the volatility obtained in point 2 - since it is based on a daily value - multiplying the standard deviation by the square root of the average number of observations per year.

The volatility recorded in the analysed period, as shown in Table 3, was much higher in the case of Ethereum (122%) and Bitcoin (74%) than in the evolution of the price of gold (18%) or of the S&P 500 index (15%). This confirms the high volatility of the cryptocurrencies in comparison with other assets. While it is true that volatility can lead to high gains in a brief period, it is also true that it can also lead to high losses.

The price of cryptocurrencies has recently been directly influenced by several shocks, some of which are related to measures taken by

Table 3. Volatility in the period between August 7, 2015, and January 15, 2023.

	Standard Deviation	Volatility
Gold	0,012	18%
S&P 500	0,008	15%
Bitcoin	0,039	74%
Ethereum	0,064	122%

Source: Own calculations.

governments. For example, Bitcoin's legalization in Japan led to its recovery while China's ban on Initial Coin Offerings (ICO) led to its devaluation (Navickas et al., 2018).

9. Investment in Cryptocurrencies

9.1 Investment in Projects through Initial Coin Offering

Cryptocurrencies (e.g., Ethereum) have allowed virtual fundraising for investment in projects, in particular start-ups, and project development through ICO.

An ICO is an innovative way of obtaining financing (fundraising) in which a project, company or individual issues digital coins and puts them up for sale in exchange for coins (digital or conventional), goods and services, or a percentage of future profits. ICO sells tokens or other cryptocurrency-based assets to investors in exchange for funding.

Several projects have come up for this purpose, such as those available on Crypto Potato (https://cryptopotato.com/ico-list/) or ICO Drops (https://icodrops.com/category/active-ico/), and others have already ended (https://icodrops.com/category/ended-ico/).

Although it appears as a way to get funding for start-ups, an activity that could initially bring benefits in economic terms, several entities (such as the European Securities and Markets Authority - ESMA) and countries (such as China - https://www.cnbc.com/2017/09/04/chinese-icos-china-bans-fundraising-through-initial-coin-offerings-report-says.html) have been warning for the high risk of ICO.

ESMA (2017a) has pointed out that companies involved in ICO, when developing regulated activities (e.g., when they qualify as financial instruments), are obliged to comply with EU legislation, stating that depending on how they are structured, they may have to comply with the following directives:

- The Prospectus Directive, that aims to "ensure that adequate information is provided to investors by companies when raising capital in the EU";
- The Markets in Financial Instruments Directive, which aims to "create a single market for investment services and activities and to ensure a high degree of harmonised protection for investors in financial instruments";
- The Alternative Investment Fund Managers Directive, which "lays down the rules for the authorisation, ongoing operation and transparency of the managers of alternative investment funds which manage and/or market alternative investment funds" in the EU;

- The Anti-Money Laundering Directive, that "prohibits money laundering and terrorist financing".

ESMA (2017b) alerts to the following risks associated with ICO:

- ICO is particularly vulnerable to illegal activities and fraud because it is an unregulated space;
- There is a risk of loss of all capital invested, as projects are at an early stage, and the guarantee of earning is very low;
- The possibility of exchange for conventional currency is reduced;
- There is a great volatility in the price of virtual currencies;
- The information provided is inadequate, focusing on the potential benefits but without much information about the risks;
- Blockchain technology flaws are still unknown but can be used to hack cryptocurrencies.

In China, new projects for obtaining funding through ICO were banned and the financial sector regulator was asked to inspect ICO's main platforms. However, this measure proved ineffective as ICO activity continued to intensify (https://bravenewcoin.com/insights/china-ico-ban-proving-ineffective). China officially banned cryptocurrencies in September 2021 (https://www.weforum.org/agenda/2022/01/what-s-behind-china-s-cryptocurrency-ban/), and it is not the only country to do so (https://fortune.com/2022/01/04/crypto-banned-china-other-countries/).

9.2 *Investment vs Speculation*

As previously referred, the cryptocurrencies have a remarkably high volatility which makes them unsuitable either as a means of payment or as a reserve of value.

Even so, the increase in the value of cryptocurrencies has attracted many users with the aim of investing and getting money quickly.

Associated with the expectation of high income through cryptocurrencies is the possibility cryptocurrencies resembling a Ponzi scheme (Krugman, 2018), where the incomes of the earlier investors are paid with the capital of new investors.[13] In this case, it is only possible to recover the funds in commercial currency if there are users interested in

[13] "A Ponzi scheme is an investment fraud that involves the payment of purported returns to existing investors from funds contributed by new investors. Ponzi scheme organizers often solicit new investors by promising to invest funds in opportunities claimed to generate high returns with little or no risk. In many Ponzi schemes, the fraudsters focus on attracting new money to make promised payments to earlier-stage investors and to use for personal expenses, instead of engaging in any legitimate investment activity." US Securities and Exchange Commission (https://www.sec.gov/fast-answers/answersponzihtm.html).

buying the cryptocurrencies, that is, if there are new users entering the system.

The fact that the cryptocurrencies do not disclose essential information, such as the risks to which investors are exposed, increases the risk that the latter will suffer financial losses.

Another significant risk is that even when investors are aware of this framework, they might still take the risk in the perspective of being able to take out the money in time.

If the decision is to invest in the acquisition of cryptocurrencies, common sense advises a diversification of investment (not only in several cryptocurrencies but also in another type of assets) to reduce the risk.

10. Lack of Regulation

The cryptocurrencies, by nature, are not subject to any regulation, that is, there is no appropriate legal basis for the operation of digital money. One of the main differences between cryptocurrencies and traditional money is that they are not regulated by a central authority. Cryptocurrencies are decentralized, meaning they are not controlled by any government, central bank, or other financial institution. This allows for some independence from public authorities and makes them attractive to some people because they offer a level of anonymity and are not subject to the same regulations as traditional currencies. However, cryptocurrencies face great legal uncertainty and risk of fraud because of this same lack of regulation.

Additionally, a large amount of money in cryptocurrencies takes away from central banks an important monetary policy instrument, undermining financial stability. Cryptocurrencies, thus, represent a major challenge for central banks, with studies advising the creation of cryptocurrencies by central banks as a viable way to deal with the problem (EBC, 2019).

On the other hand, there is no clear definition of the rights and obligations of the parties involved. In fact, users are unaware of the reliability of the remaining users and the people with whom they are doing business. The lack of information and regulation can be conducive to criminal activity. As such, the regulation of cryptocurrencies is a challenge for public authorities since it is difficult to identify the users and the transactions they perform. For this very reason, and because the information is spread by all users, there is no central point that can be turned off, a government or central bank can hardly prohibit a cryptocurrency.

Krugman (2018) points out that the cryptocurrencies are coated with a certain "technological mysticism" focusing on a "libertarian ideology" which argues the risk that governments control all money but does not take into account the risks of hacking to steal crypto-coins on a large scale.

10.1 Risks for Consumers

In recent years, concerns about the use cryptocurrencies as exchange currency were raised by entities such as the European Securities and Markets Authority, the European Banking Authority or the European Insurance and Occupational Pensions Authority (European Supervisory Authorities, 2018).

These warnings are based on the idea that cryptocurrencies are highly risky, highly speculative, have no tangible assets supporting them and are not regulated under law (including EU law), thus not offering consumer protection.

The European Supervisory Authorities (2018) highlights as the main risks the extreme volatility and risk of a bubble, the absence of consumer protection, the absence of exit options since there is a risk of not being able to switch to a conventional currency, the lack of transparency in price formation, operational problems that hinder the functioning of the system, lack of consumer information, and the fact that it is not an appropriate investment application (not only in the short term but also in the long term regarding savings for retirement).

It is also to consider the lack of guarantees and the lack of a compensation policy in case of fraud.

10.2 Income and Taxes

Money circulating outside the financial system is out of the control of the authorities, allowing a tax-exempt parallel economy. Trading with cryptocurrencies may represent a taxable event, but it is not controlled.

At the income level, there is a risk of part of the remuneration (above the minimum wage, for example) being paid in crypto-coins in order to avoid paying taxes and social security.

Regarding social security contributions and tax evasion, this situation puts the collection of revenues at risk, jeopardizing the functioning of states in areas so essential as pension payments, social support, health care, justice, security, or defence.

Regarding personal income, we should consider the same risks as mentioned above regarding consumers and investors, in particular lack of regulation, high volatility, lack of exit options and lack of compensation policy in case of fraud.

10.3 Money Laundering and Financing of Terrorist Activities

Converting money into cryptocurrency could also be a way to launder money and to finance terrorism. In fact, unlike conventional currencies, controlled by a central and regulated authority, cryptocurrencies have a high potential for use in illegal activities.

For example, "dirty money" can be exchanged successively for different cryptocurrencies, allowing not only the non-traceability of the origin but also its use in the purchase of goods and services or even in the cash back for conventional money to finance terrorist activities (European Banking Authority, 2016).

11. Energy Consumption and Environmental Impact

As previously mentioned, in transactions using the blockchain users do not need to trust each other but only in the code that serves as the basis for the network. This requires that for a transaction to be considered valid, the amount must be in the possession of those who use the currency and users must confirm that the transaction complies with this rule. For this control, the blockchain uses the "proof-of-work" in which the next valid block is the one that is sent by the first miner that produces a valid block. This miner is then rewarded. All other users are informed; they confirm the validity, and discard the blocks they were working on, restarting the process.

With special emphasis on the largest cryptocurrencies, the so-called "proof-of-work algorithm" aimed at ensuring system reliability consumes a large amount of resources - computing capacity, electrical energy and computer waste.

In this section, the analysis will be based on data from Digiconomist, "a platform that is dedicated to exposing the unintended consequences of digital trends, typically from an economic perspective". This platform estimates the impact of the two main cryptocurrencies: Bitcoin[14] and Ethereum but we will focus on the first one.

In terms of electricity consumption, Bitcoin, the most used cryptocurrency, currently uses more than 78 TWh per year, as illustrated in Figure 10.

The amount, as outlined in Figure 11, is close to that consumed by Chile and above the consumption of many other countries, resulting in an annual carbon footprint of 43,61 Mt of CO_2 per year.

According to Vries et al. (2022), as depicted in Figure 12, the share of renewables used to power the Bitcoin network decreased significantly after a mining crackdown in China during the Spring of 2021. Miners previously had access to renewables in China but lost it when they moved to countries such as the United States and Kazakhstan, where coal- and gas-based electricity is now primarily used, resulting in an increase in the

[14] In this section, whenever Bitcoin is mentioned, it refers only to the sum of the information relating to Bitcoin and Bitcoin Cash.

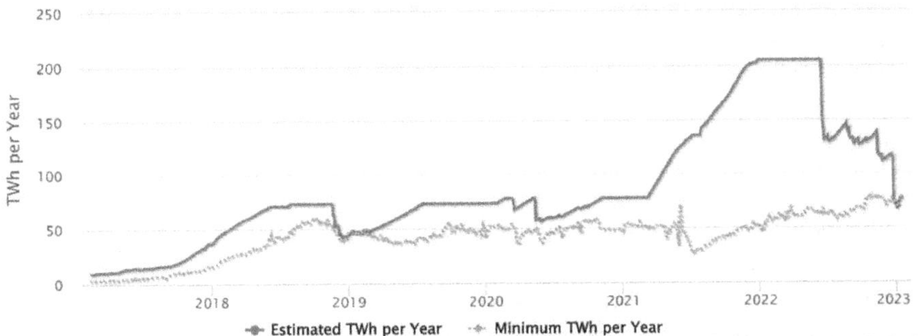

Figure 10. Bitcoin energy consumption index chart (January 19, 2017–February 4, 2023).
Source: www.bitcoinenergyconsumption.com.

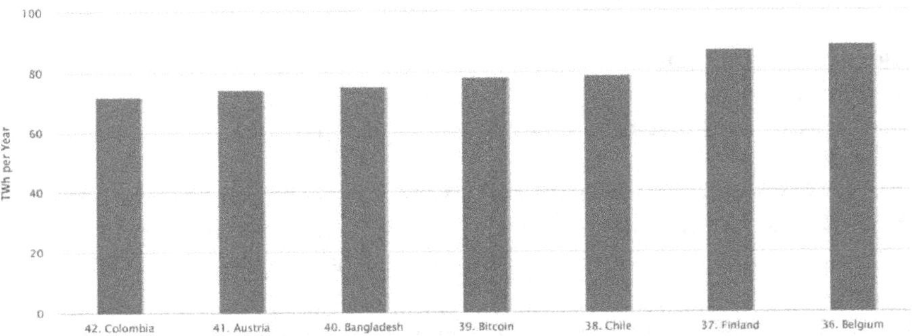

Figure 11. Energy consumption by Country (Annualised TWh).
Source: www.bitcoinenergyconsumption.com.

carbon intensity of electricity used for Bitcoin mining. The average carbon intensity of electricity consumed by the Bitcoin network, according to the referred authors, may have increased about 16.6% from 2020 to 2021, reaching 557.76 gCO_2/kWh in August 2021.

This impact increases as the number of cryptocurrency transactions increases (and vice versa) and could threaten global efforts to reduce CO_2 emissions. It should be noted that recent developments in bitcoins have resulted in a decrease in these values since June 2022.

Additionally, as illustrated in Figure 13, the pressure on computers leads to the rapid devaluation of less efficient equipment, causing electronic waste to reach more than 40 kt per year.

Based on data from Figure 14, Bitcoin's high electronic waste footprint reaches 449 grams per transaction, slightly lower than one Apple iPad (490 grams) but higher than one Apple iPhone 12 (164 grams) and much higher than the e-waste of 10,000 VISA transactions (40 grams).

Recently, Vries et al. (2021) analysed Bitcoin's electronic waste and concluded that "it adds up to 30.7 metric kilotons annually, per May 2021"

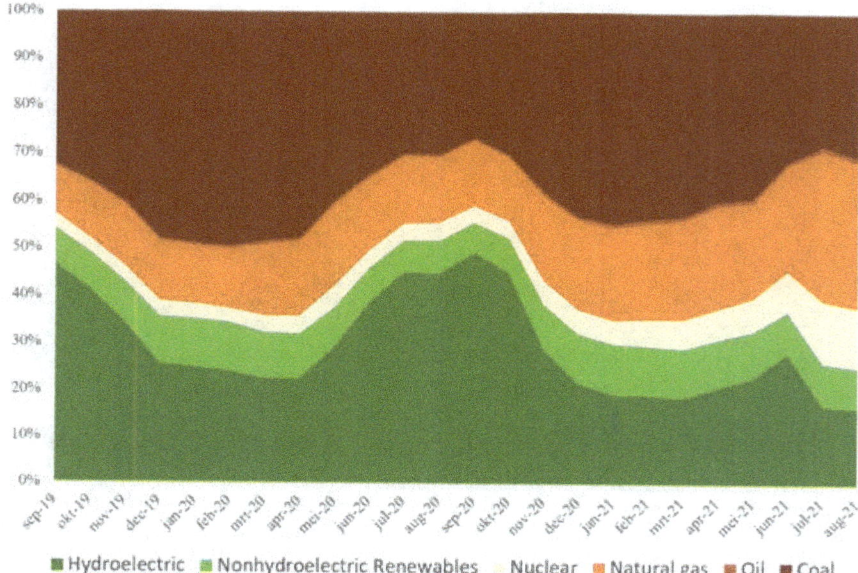

Figure 12. Electricity mix of the Bitcoin network (September 2019–August 2021).
Source: Digiconomist (https://digiconomist.net/bitcoin-energy-consumption) based on Vries et al. (2022).

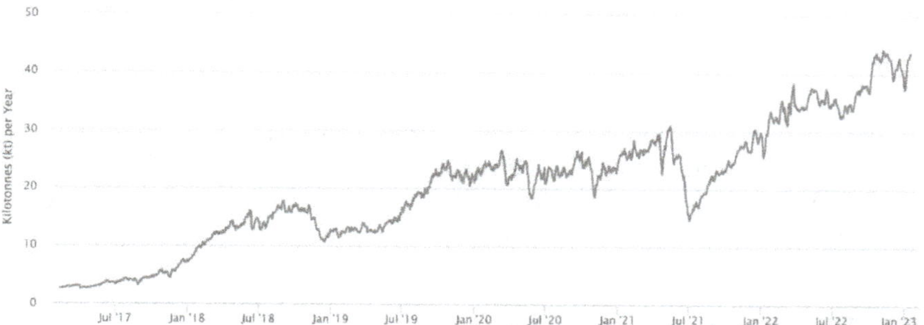

Figure 13. Bitcoin electronic waste generation.
Source: www.bitcoinelectronicwaste.com.

and that "at peak Bitcoin price levels seen early in 2021, the annual amount of e-waste may grow beyond 64.4 metric kilotons in the midterm".

Considering annualized information for Bitcoin, the estimated global mining costs represent almost 75% of the global mining revenues.

The second largest cryptocurrency, Ethereum, due to its size has a much lower consumption than Bitcoin. Still, in annualized terms, estimated global mining costs represent more than 77% of global mining revenues.

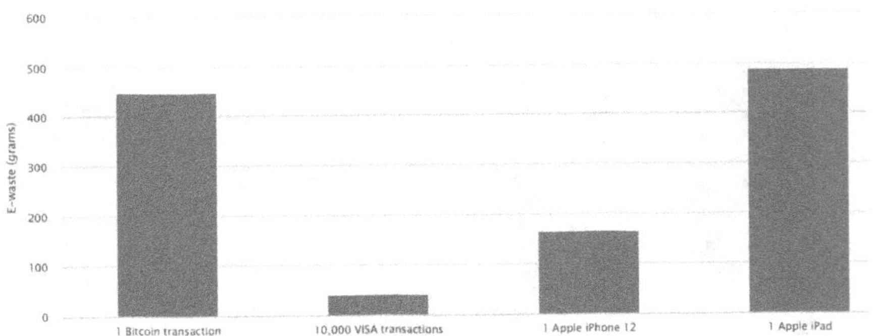

Figure 14. Electronic waste footprint (grams).
Source: www.bitcoinelectronicwaste.com

As mentioned before, one of the main immediate consequences of the cryptocurrencies is the consumption of computational resources, both in the consumption of electricity and in the degradation of the performance of computers. According to Microsoft (2019), continuous crypto-coin mining can lead to performance problems on computers and excessive CPU utilization can even damage computers. The European Banking Authority (2019) also warns for the energy consumption of cryptocurrencies and for the need for a cost-benefit analysis that considers the environmental protection and the sustainable development of the energy sector. In this regard, this technological revolution also entails environmental and social costs.

Cagnin et al. (2021) consider that, although "Blockchain has the potential to revolutionise the way we share information and carry out transactions online", issues remain, namely those regarding "high-energy consumption". Even though, the same authors consider that "it is believed that blockchain will lead the way to a sustainable global economy".

12. Security Issues and Attacks

The history of cryptocurrencies has not been exempt from attacks that seek to exploit the pseudonymity of users in current networks, trying to influence the behaviour of the network for malicious purposes. Although proof-of-work makes the mining process more resilient to some of these threats, it is not impossible that attacks like the ones referred below take place. Generally, the attacks are quite sophisticated and require knowledge of the technology and the network, but it's important to be aware of these risks and take steps to protect your assets accordingly.

12.1 Access to Clipboard through Malware

Through phishing/malware, by exploiting flaws in security, attackers may access wallets and exchange accounts to steal cryptocurrencies. One way to jeopardize the ownership of cryptocurrencies and allow them to be stolen, is via the clipboard. Using phishing/malware, attackers may, for instance, access the clipboard (copying/paste). Malware can also covertly cause payment to be directed to people other than those who should be receiving it, without needing to crack passwords or even copying keys.

12.2 Double Spend Attack or 51% Attack

In centralized systems, the banks guarantee the validity of operations. In the case of cryptocurrencies based on blockchain, the decision results from the votes of the peers, and the protocol becomes a kind of governance that seeks to replace the existence of a bank system.

To add new blocks to the blockchain users vote and reach a consensus. This proof-of-work associates voting with hashing power and its correct functioning presupposes that most of the computational power is associated with honest users and that this majority will be able to mine faster than a malicious user.

However, if a malicious entity can control more than 50% of computational power, it can double spend the same crypto-coins in the blockchain. Sometimes the goal is not to spend twice but to destabilize and discredit a cryptocurrency by affecting its integrity.

Nevertheless, it is important to note that this type of attack is extremely difficult to execute and the probability of success of an attack is smaller the larger the network, as this would require an exorbitant expenditure with hardware, storage space and electricity to compete with the rest of the network. In fact, although this type of attack has already occurred in the past in smaller blockchains (e.g., Krypton, 2016; Verge, 2018; MonaCoin, 2018), it has never occurred with Bitcoin blockchain (although not impossible).

12.3 Distributed Denial-of-Service

A distributed denial-of-service (DDoS) attack is the most common network attack and aims to stop normal traffic from a server or network flooding the target with excessive internet traffic. By compromising multiple computers, servers, or networks, it turns them into sources of traffic to carry out the attack. These attacks can be carried out at no cost to attackers.

In the specific case of cryptocurrencies, this attack seeks to deny service or drive away honest users, increasing the hash rate of malicious miners. For example, if an honest miner receives a high number of transactions from many customers, he will start discarding new orders, even from honest customers. In this way, the attackers seek to obtain most of the hash power.

Because of the decentralized nature of networks based on blockchain technology and proof-of-work, such an attack will have a very limited effect unless a large-scale attack is launched.

12.4 Sybil Attack

Similarly, the Sybil Attack floods the network with zero-power nodes. In a system where users are not identifiable (even in Bitcoin where there is no strong anonymity) and in which there is no central authority that confers the identity of the participants, it is possible to create many false entities to which multiple votes correspond.

In this way, the attackers give the idea that there are a lot of different participants when in fact they are pseudo-identities controlled by the attackers.

This operation will allow the attacker, for example, to perform "double spend" (as previously seen).

12.5 Forking

Forking occurs when different miners create different blocks to place at the same point in the blockchain, creating several versions of the transaction history. The solution resulting from the protocol implies that the miners agree which valid chain will be the one on which they will continue to build blocks.

Forking may happen unintentionally but may also occur intentionally with the purpose of making changes to the protocol to validate transactions that were previously considered invalid or to attack specific addresses by preventing them from transacting which may constitute a form of blackmail in which the attacker will fork all transactions of a particular user, invalidating them until he pays a certain amount.

12.6 Malicious Cryptocurrency Miners

According to Microsoft (2019), the mining of crypto-coins can be very profitable but it takes huge computing power that implies a great expense of resources. For this reason, attackers have increasingly turned to malware[15] to gain access to victims' computers that they use to increase

[15] On this regard, Barros (2018) refers that Portugal has one of the highest rates of malware incidence among EU28 countries (9th position). According to Microsoft (2017), in March 2017 malware was found in 8.3% of computers in Portugal, compared to a worldwide incidence rate of 7.8%.

their own computing power. In this situation, the victims do not realize that they are being used because the mining occurs in the background. The only effect being eventually visible is the reduction of the performance of computers. For this reason, these attacks can last for long time without being detected or fought.

Microsoft (2019) also notes that many of the malicious mining processes do not need to be installed and are based only on JavaScript-enabled browsers (through, for example, introducing the mining code into supposedly credible sites), allowing attackers to mine crypto-coins without the consent of the users. Again, the user is not aware that the computer is being used and computer degradation may occur.

12.7 *Chain Hopping*

In chain hopping attacks, the hackers move funds between different blockchain networks to take advantage of differences in security or regulatory oversight.

The attacker will typically transfer funds from a blockchain network that has a higher level of security or regulatory oversight to one that has a lower level. This allows the attacker to potentially evade detection or prosecution and can also result in the loss of funds for the victim.

Chain hopping can be accomplished by using a technique called "atomic swaps" (https://corporatefinanceinstitute.com/resources/cryptocurrency/atomic-swaps/), which allows for the exchange of one cryptocurrency for another without the need for a centralized exchange. This makes it possible for a hacker to move funds between different blockchains in a highly automated and efficient manner.

12.8 *Ransomware Attacks*

Also related to cryptocurrencies, we may consider the ransomware attacks, which is a type of attack in which a hacker encrypts a user's files and demands payment in cryptocurrency to restore the access.

These attacks can be particularly devastating for businesses, as they can result in the loss of important data and disrupt operations.

The attackers typically use phishing or social engineering tactics to gain access to a victim's network, and then use malware to encrypt the files. Once the files have been encrypted, the attacker will typically display a message demanding payment to restore access to the files. The payment is usually requested in a cryptocurrency such as Bitcoin, as it allows for anonymity and is difficult to trace.

13. Credibility of Cryptocurrencies in the Early Years and in Recent Years

The beginning of Bitcoin was marked, by two major scandals that shook the nascent cryptocurrency market.

One of these situations regards the Silk Road case. The website was launched in February 2011, becoming the largest anonymous online market in the world, known for its large collection of illicit drugs, reaching some 960,000 registered users (Pagliery, 2013).

The website operated through Darknet, which used the Tor network, and transactions were paid in Bitcoin, ensuring the anonymity of buyers and sellers. By August 2012, it was estimated that annual site sales would reach 22 million USD (Greenberg, 2012). In October 2013, the Federal Bureau of Investigation closed the website Silk Road and arrested the alleged site owner, Ross William Ulbricht.[16]

This incident had a significant impact on the public perception of Bitcoin and cryptocurrencies more broadly. Many people saw the association with illegal activity as seriously negative and it raised concerns about the ability of cryptocurrencies to facilitate criminal activity. However, it is important to note that the use of Bitcoin for illegal purposes is not unique to the Silk Road case, and it is not an inherent characteristic of cryptocurrencies.

The second major scandal that affected the credibility of cryptocurrencies was the collapse of Mt. Gox, a Bitcoin trading platform. In 2014, Bitcoin was again associated with a police case, relating to Mt. Gox. Mt. Gox was a Bitcoin trading platform launched in 2010 and in 2013 operated about 70% of all Bitcoin transactions (Frunza, 2015).

In February 2014, Mt. Gox abruptly shut its website down and filed for bankruptcy protection (Abram et al., 2014), and in April of the same year began to wind up proceedings (Mochizuki et al., 2014). Mt. Gox then stated that bitcoins worth approximately 450 million USD were missing and probably had been stolen (Abram et al., 2014).

This incident raised serious questions about the security and stability of Bitcoin and other cryptocurrencies, and it had a significant impact on the market.

Over the years, there have been several high-profile scandals involving cryptocurrencies.

Similar to Mt. Gox, there were other cases of cryptocurrency exchanges being hacked or suffering other security breaches, resulting in the theft

[16] https://web.archive.org/web/20140220003018/https://www.cs.columbia.edu/~smb/UlbrichtCriminalComplaint.pdf.

of millions of dollars' worth of digital assets. We highlight the hack of Bitfinex (2016) and the hack of Coincheck (2018).

These incidents have raised concerns about the security of cryptocurrency exchanges and have contributed to the overall volatility of the market.

Although some ICOs have been successful and have generated significant returns for investors, others have been fraudulent or have failed to deliver on their promises, leading to regulatory scrutiny and increased caution among investors.

Despite these challenges, cryptocurrencies have continued to grow and evolve over the years, proposing to be a viable alternative to traditional fiat currencies.

The year 2022 was critical for crypto industry, with prices dropping significantly and pushing some companies to bankruptcy. The major examples, according to Knauth (2022) were:

- The crypto exchange FTX – FTX was a Bahamas-based exchange that had a valuation of $32 billion at the start of the year but went bankrupt after a series of setbacks including the failure of a potential merger with rival exchange Binance and allegations of mismanagement by its founder;
- The crypto lender BlockFi - BlockFi, a crypto lender, filed for Chapter 11 bankruptcy a few weeks after FTX's collapse, having relied on a $400 million credit facility from FTX to stay afloat;
- The crypto hedge fund Three Arrows Capital (3AC) - Three Arrows Capital, a crypto hedge fund based in Singapore, filed for bankruptcy in the British Virgin Islands in June after the collapse of cryptocurrencies Luna and TerraUSD wiped out $42 billion in investor value;
- The crypto lender Voyager - Voyager Digital, a New Jersey-based crypto lender, filed for bankruptcy in the United States in July after defaulting on a $650 million crypto loan;
- The crypto lender Celsius Network - Celsius Network, another crypto lender, also filed for bankruptcy in the United States in July amid disputes over fraud investigations, customer privacy, and spending on a new bitcoin mining facility.

According to Rosenberg (2022), crypto companies that declare bankruptcy can leave investors with limited recourse to recover their funds. Cryptocurrencies are not insured by the central banks like traditional banks, meaning that if a crypto exchange goes out of business, no government agency will compensate investors. In the event of bankruptcy, secured creditors are normally paid first, followed by unsecured creditors, and investors are last in line to recover their assets. The bankruptcy process can be complex and time-consuming, and investors may not receive their full investment back or may receive it in the form of company stock rather than cash. It is important for crypto investors to consider the risks of the exchanges or lending platforms they use and to diversify their holdings to mitigate potential losses.

14. New Digital Currencies – Stable Coins and Central Bank Digital Currencies

Recently, new forms of cryptocurrencies appeared, the called stablecoins, which are grounded on reserve assets like a currency (Euro or Dollar, for instance), other assets (gold, for example) or even in other cryptocurrencies with the objective of controlling volatility. These stablecoins, as other cryptocurrencies, are not issued by central banks. Institutions such as the European Central Bank or the Federal Reserve Board have been worried about the impact of stablecoins on financial stability and on monetary policy.[17]

Tether, USD Coin, Binance (https://www.benzinga.com/money/best-stablecoins-and-4-types-of-stablecoins) are nowadays some of the most important stablecoins and among the top five cryptocurrencies.

Another example is Diem (previously called Libra), the cryptocurrency created by Facebook based on open-sourced Libra Blockchain. The initial white paper was published in June 2019 but due to the evolution the project a new version was released in April 2020[18] and ended in January 2022.[19]

New forms of digital currency that do not constitute cryptocurrencies are currently also being implemented and may be relevant in the future in terms of means of payment available to citizens and businesses – the central banks digital currencies. Central banks in several countries are seeking to modernise their financial systems to face the challenges posed to the financial system by cryptocurrencies.

An example of these digital currencies is the Digital Yuan, created by the Chinese central bank (The People's Bank of China), which aims to become an alternative to coins and banknotes in circulation (https://www.china-briefing.com/news/china-launches-digital-yuan-app-what-you-need-to-know/). According to a senior Chinese central bank official quoted

---

[17] https://www.cnbc.com/2021/08/18/the-fed-is-worried-the-rise-of-stablecoins-could-impact-financial-stability.html and https://www.ecb.europa.eu/press/key/date/2020/html/ecb.sp201104~7908460f0d.en.html.

[18] https://www.diem.com/en-us/white-paper/.

[19] https://corporatefinanceinstitute.com/resources/cryptocurrency/libra-cryptocurrency/.

by Reuters,[20] around 140 million people opened "wallets" for the new digital yuan in October and used it for transactions worth approximately USD 9.7 billion.

Also, the European Union, through the Eurosystem (European Central Bank and national central banks) intends to create the Digital Euro (European Central Bank, 2020), which will continue to represent euros and can be used without any cost by citizens and companies, just like the current coins and banknotes, but constituting a safe and reliable means of payment for electronic payments thanks to the existence of a legal framework that shall guarantee their security and protection.[21] The Eurosystem is studying the design of the Digital Euro to provide a means of payment useful for European citizens and enterprises while protecting the financial stability and the monetary policy. Additionally, the development of the European digital currency is part of a response not only to the challenges posed by cryptocurrencies but also to the risks that the Chinese digital currency poses to the role of the Euro in international terms.

The US Federal Reserve has also been discussing and considering the need to create a Digital Dollar to maintain the importance of the American dollar in the international financial system. However, so far, there has been no decision to introduce a digital currency into the US financial system,[22] and the advantages and challenges of moving forward with such a measure in the context of payment systems are still being considered.

When evaluating payment methods, it is vital to consider both their advantages and drawbacks. Paramount among considerations are security, ensuring transactions are safe, user-friendliness for ease of use and affordability. As for new digital currencies, they will not be entirely anonymous. It is essential to exercise control over their utilization and address potential challenges through proper regulation.

In the process of creation of digital currencies by central banks there are non-linearities that will lead to specificities in each central bank that result from changes in geopolitical relations and perceptions about the advantages and disadvantages of this type of currency for financial systems.

---

[20] https://www.reuters.com/technology/95-billion-spent-using-chinese-central-banks-digital-currency-official-2021-11-03/.

[21] https://www.ecb.europa.eu/paym/digital_euro/report/html/index.en.html.

[22] https://www.federalreserve.gov/faqs/what-is-a-central-bank-digital-currency.htm.

15. Final Remarks

Cryptocurrencies have gained great visibility and adhesion in a short time but they are not exempt from risks. This study identified the main advantages and risks associated with this new form of digital money.

Among the main advantages are the ease and speed of transaction, the privacy (pseudonymity, for example in the case of Bitcoin, which does not correspond to total anonymization), the system is available 24 hours a day, the costs associated with transactions are low and the crypto-coins in circulation are limited (which avoids inflationary processes arising from the issuance of currency).

However, several risks associated with cryptocurrencies have been identified: they are highly volatile (which makes them highly speculative), privacy attracts criminal activity, lack of central and supervisory authority (namely undermining monetary policy), difficulty in identifying users (which makes them vulnerable to hacking, terrorism and money laundering), there is no system to resolve the dispute between the parties (which is not possible because there is no central authority) and there is no deposit guarantee system.

Cryptocurrencies are only accepted by a small number of traders, which makes their use more difficult. It is even illegal to trade in some countries, in Bitcoin, for example. It is also highlighted that the crypto-coins are lost if the owner loses the private keys.

This study also emphasizes the applications associated with blockchain. It seems possible that blockchain will revolutionize financial and economic infrastructure - moving from the idea of decentralized crypto-concept to the idea of a decentralized world. However, blockchain is a technology in development and the success of the application to other sectors is still ongoing. Its use is still at an early stage and there is some scepticism about the results it will provide.

In the future, several challenges arise, not only in relation to blockchain but in relation to cryptocurrencies, that require technological evolution. Regarding the mentioned attacks, there are studies that seek to improve the safety of the cryptocurrencies, so that the existing blockchain system can be improved in the future.

To face this evolution, in the future we can expect the creation of digital currencies by central banks to combat some of the risks associated with cryptocurrencies, particularly regarding the lack of any guarantee or legal protection for cryptoactives.

Cryptocurrencies may be passing through a period of some turbulence, but it is impossible to predict their future. It will depend, of course, on the capacity to respond to the challenges identified in this paper, on the way financial markets react to these alternatives, on geopolitical strategies, and on the measures by central banks and financial systems to address the threats cryptocurrencies represent to the financial stability and monetary policy.

References

Abrams, R., Goldstein, M. and Tabuchi, H. (2014). Erosion of Faith Was Death Knell for Mt. Gox.

Aggarwal, D., Brennen, G., Lee, T., Santha, M. and Tomamichel, M. (2017). Quantum attacks on Bitcoin, and how to protect against them.

Back, A. (2002). Hashcash - A Denial of Service Counter - Measure.

Baldi, M. and Franco, C. (2017). A trusted cryptocurrency scheme for secure and verifiable digital transactions. First Monday, 22(11).

Bansal, L. (2017). Blockchain: A New Type of Internet - How Interconnection is Powering Global Cryptocurrency.

Barros, G.O. (2018). A Cibersegurança em Portugal. Tema Económico no. 56, Gabinete de Estratégia e Estudos do Ministério da Economia.

Ben-sasson, E., Chiesa, A., Garman, C., Green, M., Miers, I., Tromer, E. and Virza, M. (2014). Zerocash: Decentralized Anonymous Payments from Bitcoin, 459–474.

BerkeleyX/edX. (2018). Bitcoin and Cryptocurrencies. Blockchain Fundamentals Program.

Biais, B., Bisière, C., Bouvard, M. and Casamatta, C. (2018). The blockchain folk theorem. Toulouse School of Economics (TSE), Working Papers No., 17–817.

Böhme, R., Christin, N., Edelman, B. and Moore, T. (2015). Bitcoin: Economics, technology, and governance. Journal of Economic Perspectives, 29(2): 213–38.

Bonneau, J., Narayanan, A., Miller, A., Clark, J., Kroll, J.A. and Felten, E.W. (2014). Mixcoin - Anonymity for Bitcoin with accountable mixes. Part of the Lecture Notes in Computer Science book series (LNCS, volume 8437).

Bonneau, J., A. Miller, J. Clark, A. Narayanan, J.A. Kroll and E.W. Felten, (2015). SoK: research perspectives and challenges for bitcoin and cryptocurrencies. IEEE Symposium on Security and Privacy, San Jose, CA, USA, pp. 104–121, doi: 10.1109/SP.2015.14.

Bouoiyour, J. and Selmi, R. (2017a). Are Trump and Bitcoin Good Partners?

Bouoiyour, J. and Selmi, R. (2017b). Ether: Bitcoin's competitor or ally? Working Papers hal-01567277, HAL.

Buterin, V. (2014). Ethereum White Paper.

Cagnin, C., Muench, S., Scapolo, F., Störmer, E. and Vesnic-Alujevic, L. (2021). Shaping and securing the EU's open strategic autonomy by 2040 and beyond. Publications Office of the European Union.

Caporale, G.M., Gil-Alana, L. and Plastun, A. (2017). Persistence in the cryptocurrency market. DIW Berlin Discussion Paper No. 1703.

Chatzopoulos, D., Ahmadi, M., Kosta, S. and Hui, P. (2017). FlopCoin: A cryptocurrency for computation offloading. IEEE Transactions on Mobile Computing.

Chaum, D. (1985). Security without identification: Transaction systems to make big brother obsolete. Comm. ACM 28: 10.

Chaum, D. (1988). Privacy protected payments: Unconditional payer and/or payee untraceability. pp. 69–93. *In*: Chaum, D. and Schaumuller-Bichl, I. (eds.). Smartcard 2000. Amsterdam, North Holland.

Chaum, D., Fiat, A. and Naor, M. (1990). Untraceable electronic cash. *In*: Goldwasser, S. (ed.). Advances in Cryptology - CRYPTO' 88. CRYPTO 1988. Lecture Notes in Computer Science, vol 403. Springer, New York, NY.

Conley, J.P. (2017). Blockchain Cryptocurrency Backed with Full Faith and Credit. Vanderbilt University Department of Economics Working Papers 17-00007, Vanderbilt University Department of Economics.

Conti, M., Kumar, E., Lal, C. and Ruj, S. (2017). A survey on security and privacy issues of bitcoin. Published in IEEE Communications Surveys & Tutorials, 20(4).

Dagher, G.G., Bünz, B., Bonneau, J., Clark, J. and Boneh, D. (2015). Provisions: Privacy-preserving proofs of solvency for Bitcoin exchanges. In CCS 2015—Proceedings of the 22nd ACM SIGSAC Conference on Computer and Communications Security, Vol. 2015-October, pp. 720–731, Association for Computing Machinery.

Dai, W. (1998). B-Money, an anonymous, distributed electronic cash system.

Davradakis, E. and Santos, R. (2019). Blockchain, FinTechs and their relevance for international financial institutions. EIB Working Papers, No. 2019/01, European Investment Bank (EIB).

Dwork, C. and Naor, M. (1993). Pricing via processing or combatting junk mail. *In*: Brickell, E.F. (ed.). Advances in Cryptology—CRYPTO' 92. CRYPTO 1992. Lecture Notes in Computer Science, vol 740. Springer, Berlin, Heidelberg.

Elbahrawy, A., Alessandretti, L., Kandler, A., Pastor-Satorras, R. and Baronchelli, A. (2017). Evolutionary dynamics of the cryptocurrency market. Royal Society Open Science, 4.

European Banking Authority. (2014). EBA Opinion on 'virtual currencies'. EBA/Op/2014/08.

European Banking Authority. (2016). Opinion of the European Banking Authority on the EU Commission's proposal to bring Virtual Currencies into the scope of Directive (EU) 2015/849 (4AMLD). EBA-Op-2016-07.

European Banking Authority. (2019). Report with Advice for the European Commission on Crypto-assets.

European Central Bank. (2012). Virtual Currency Schemes.

European Central Bank. (2015). Virtual Currency Schemes—A Further Analysis.

European Central Bank. (2018). What is Bitcoin?

European Central Bank. (2019). Crypto-Assets: Implications for financial stability, monetary policy, and payments and market infrastructures. ECB Occasional Paper Series No. 223, ECB Crypto-Assets Task Force.

European Central Bank. (2020). Report on a Digital Euro.

European Securities and Markets Authority. (2017a). ESMA alerts firms involved in Initial Coin Offerings (ICOs) to the need to meet relevant regulatory requirements. ESMA, 50-157-828.

European Securities and Markets Authority. (2017b). ESMA alerts investors to the high risks of Initial Coin Offerings (ICOs). ESMA, 50-157-829.

European Supervisory Authorities. (2018). ESMA, EBA and EIOPA warn Consumers on the risks of Virtual Currencies.

Eyal, I. and Sirer, E.G. (2014). Majority is not Enough: Bitcoin Mining is Vulnerable. Conference Paper, Chapter from book Financial Cryptography and Data Security: 18th International Conference, FC 2014, Christ Church, Barbados.

Fahmy, S.F. (2018). Blockchain and its Uses.

Frunza, M. (2015). Solving Modern Crime in Financial Markets: Analytics and Case Studies.

Gainsbury, S. and Blaszczynski, A. (2017). How blockchain and cryptocurrency technology could revolutionize online gambling. Gaming Law Review, 21: 482–492.

Gandal, N., Hamrick, J.T., Moore, T. and Oberman, T. (2017). Price manipulation in the Bitcoin ecosystem. Journal of Monetary Economics.

Gerlach, J.C., Demos, G. and Sornette, D. (2018). Dissection of bitcoin's multiscale bubble history from January 2012 to February 2018. SSRN Electronic Journal.

Greenberg, A. (2012). Black Market Drug Site 'Silk Road' Booming: $22 Million In Annual Sales. Forbes.

Guo, L. and Li, X.J. (2017). Risk analysis of cryptocurrency as an alternative asset class. pp. 309–329. *In*: Härdle, W.K. et al. (eds.). Applied Quantitative Finance, Statistics and Computing.

Harvey, C.R. and Tymoigne, E. (2015). Do cryptocurrencies such as bitcoin have a future? Wall Street Journal.

Hegadekatti, K. and S.G., Y. (2016a). Examining taxation of fiat money and bitcoins vis-a-vis regulated cryptocurrencies. Published in International Finance eJournal, 8(117).

Hegadekatti, K. and S.G., Y. (2016b). Banking systems in an economy dominated by cryptocurrencies. Published In: Monetary Economics: Central Banks - Policies & Impacts eJournal , 01(81): 1–16.

Hoskinson, C. (2017). Why we are building Cardano – A Subjective Approach.

Hotz-Behofsits, C., Huber, F. and Zörner, T.O. (2018). Predicting crypto-currencies using sparse non-Gaussian state space models. Journal of Forecasting, 37: 627–640.

Jonker, N. (2017). What drives virtual currency adoption by retailers? Payments Conference 2017 Academic Paper, Joint ECB/BdI "Digital transformation of the retail payments ecosystem".

Kampl, A. (2014). Analysis of Large-Scale Bitcoin Mining Operations. White Paper, Applied Control.

Knauth, D. (2022). Factbox: Crypto companies crash into bankruptcy. Reuters.

Krogt, D. van der (2018). GARCH Modeling of Bitcoin, S&P-500 and the Dollar. Bachelor Thesis Financial Economics, Erasmus University.

Kroll, J.A., Davey, I.C. and Felten, E.W. (2013). The economics of bitcoin mining, or bitcoin in the presence of adversaries. In Proceedings of The Twelfth Workshop on the Economics of Information Security (WEIS).

Krugman, P. (2018). Bitcoin is basically a Ponzi scheme. The Seattle Times, 30 January.

Lee, D.K.C., Guo, L. and Wang, Y. (2018). Cryptocurrency: A new investment opportunity? Journal of Alternative Investments, 20(3): 16–40. Research Collection Lee Kong Chian School of Business.

Lielacher, A. (2018). The History of Bitcoin Part 1: What is Hashcash?. Cryptocyclopedia.

Lin, Q., Yan, H., Huang, Z., Chen, W., Shen, J. and Tang, Y. (2018). An ID-based linearly homomorphic signature scheme and its application in blockchain. In IEEE Access, 6: 20632–20640.

Li, X. and Wang, C.A. (2017). The technology and economic determinants of cryptocurrency exchange rates: The case of bitcoin. Decision Support Systems, 95: 49–60.

Mersch, Y. (2018). Virtual currencies ante portas. Speech by Yves Mersch, Member of the Executive Board of the ECB, at the 39th meeting of the Governor's Club Bodrum, Turkey, 14 May 2018.

Microsoft. (2019). Microsoft Security Intelligence Report. Volume 24, January–Dezember 2018.

Mochizuki, T. and Stech, K. (2014). Mt. Gox Files for Liquidation - Defunct Bitcoin Exchange Gives Up On Plan to Rebuild.

Moore, T. and Christin, N. (2013). Beware the middleman: Empirical analysis of bitcoin-exchange risk. In Financial Cryptography and Data Security, vol. 7859 of Lecture Notes in Computer Science, pp. 25–33. Springer.

Möser, M., Böhme, R. and Breuker, D. (2013). An inquiry into money laundering tools in the bitcoin ecosystem. *In*: eCrime Researchers Summit (eCRS).

Möser, M. and Böhme, R. (2015). Trends, tips, tolls: A longitudinal study of bitcoin transaction fees. 2nd Workshop on Bitcoin Research, affiliated with the 19th International Conference on Financial Cryptography and Data Security, Puerto Rico.

Nakamoto, S. (2008). Bitcoin: A Peer-to-Peer Electronic Cash System.

Narayanan, A., Bonneau, J., Felten, E., Miller, A. and Goldfeder, S. (2016). Bitcoin and Cryptocurrency Technologies: A Comprehensive Introduction. Princeton University Press.

Navickas, M., Bagdonas, I. and Ngoc Viet Nguyen, V. (2018). Predicting Bitcoin Price using Machine Learning.

Pagliery, J. (2013). FBI shuts down online drug market Silk Road. CNN Money.

Rosenberg, E. (2022). What Happens when a Crypto Exchange Goes Bankrupt? Investopedia.

Rosenfeld, M. (2011). Analysis of bitcoin pooled mining reward systems. arxiv:1112.4980v1.

Sharma, A. (2018). Digital Money- Bitcoin. In Digitalization, Chapter 30, pp. 122–124.

Shaw, C. (2018). Conditional heteroskedasticity in crypto-asset returns. Munich Personal RePEc Archive, Paper No. 90437.

Smalley, C.W. (2017). Cryptocurrency and taxes. The Tax adviser, Tax Insider, pp. 1–3.

Staley, H. (2018). Blockchain technology as a disruptor in Finance. The Business and Management Review, 9(3).

Stavroyiannis, S. (2017). Value–at-risk and expected shortfall for the major digital currencies. Article in SSRN Electronic Journal.

Taylor, M.B. (2013). Bitcoin and the age of bespoke silicon. International Conference on Compilers, Architecture and Synthesis for Embedded Systems (CASES).

Vasek, M., Thornton, M. and Moore, T. (2014). Empirical analysis of denialof-service attacks in the bitcoin ecosystem. Financial Cryptography and Data Security: FC 2014 Workshops, BITCOIN and WAHC, Springer Berlin Heidelberg, pp. 57–71.

Vasek, M. and Moore, T. (2015). There's no free lunch, even using bitcoin: tracking the popularity and profits of virtual currency scams. Financial Cryptography and Data Security. FC, 2015.

Velankar, S., Valecha, S. and Maji, S. (2018). Bitcoin price prediction using machine learning. 20th International Conference on Advanced Communications Technology (ICACT), Chuncheon-si Gangwon-do, Korea (South), pp. 144–147.

Vora, G. (2015). Cryptocurrencies: Are disruptive financial innovations here? Modern Economy, 6: 816–832.

Vries, A. and Stoll, C. (2021). Bitcoin's growing e-waste problem. Resources, Conservation and Recycling, 175.

Vries, A., Gallersdörfer, U., Klaaßen, L. and Stoll, C. (2022). Revisiting Bitcoin's Carbon Footprint.

Walter, T. and Klein, T. (2018). Exogenous drivers of cryptocurrency volatility—A mixed data sampling approach to forecasting. University of St. Gallen, School of Finance Research Paper No. 2018/19.

World Economic Forum. (2019). Globalization 4.0 - Shaping a New Global Architecture in the Age of the Fourth Industrial Revolution. White Paper.

Chapter 10

Blockchain-based P2P Carsharing towards Sustainability

*José Rui Sousa** and *António Pimenta de Brito*

1. Introduction

The issue of sustainability is not a new one, but it is clear that the last few years have brought a dynamic of not only doing the right thing, but doing it in an intelligent way. It is no longer a question of what to do, but how best to do it, and in a way that everyone can contribute. One such case involves the need to rethink the cities we live in, not only from an environmental standpoint, but also in terms of quality of life for those who live there, creating a positive feedback loop effect - more sustainable habits create more sustainable environments, which in turn encourage even more sustainable habits as quality of life improves.

Although the concept of sharing has existed for several decades (Belk, 2010), recent Web2 technology innovations have introduced the so-called Platform Economy that has allowed sharing to move from local and isolated systems to global and comprehensive systems (Voshmgir, 2020).

BRU-ISCTE/ESCAD-IP LUSO, ISCTE-Instituto Universitário de Lisboa, ESCAD-IP Luso – Universidade Lusófona.
Email: acpbo@iscte-iul.pt
* Corresponding author: ruisousa@myblockchain.pt

The sharing economy, which is based on a peer-to-peer model, plays an important role in the efficient use of limited and underutilized resources, which in turn reduces waste and promotes sustainability—including the sustainability of cities. The sharing economy fits perfectly within the scope of sustainability by acting directly on the 3 strands of sustainability—economic, environmental and societal (Akande et al., 2020). On the economic side, the sharing economy enables the reuse by others of assets that are mostly sitting idle or underutilized, inclusively for monetary compensation, but potentially less costly than owning that asset; and together with other technological innovations, it promotes new business models (Bernardi and Diamantini, 2018). From an environmental perspective, the sharing economy enables the reduction of resources and assets that would otherwise be duplicated, with direct implications for reducing the carbon footprint. On the social side, the sharing economy creates more intertwined and empowered societies, where the act of each person affects the whole, and thus, the self itself (Bernardi and Diamantini, 2018); additionally, it accompanies changes in consumer behavior regarding the ownership of goods and the need for social connection (Botsman and Rogers, 2010).

Carsharing is a sharing economy use case that addresses all three strands of sustainability perfectly. It reduces the number of cars in cities (and their overproduction), reducing the carbon footprint, without harming mobility; it promotes the development of new business models or product enhancement, not only for OEMs but also for individuals; it creates more functional and efficient societies with latent effects on the quality of life.

Web2 technology has boosted networked sharing solutions, whether B2B (business-to-business), B2C (business-to-consumer), C2C (consumer-to-consumer); Web3 technology is now boosting P2P (peer-to-peer) (Münzel et al., 2019). The concept of carsharing falls into the latter two, but P2P lacks a closer look as to the operational efficiency of the solutions architecture. Both C2C and P2P involve consumers or individuals dealing with one another. The main difference between them is that in the case of C2C, there is a company or third party that establishes the connection point between the two, while in the case of P2P the exchanges are made directly from individual to individual, without any third party involved in the process. However, in a social network of actors unknown to each other, and without any central party dictating the rules of the game, how can the trustworthiness, not maliciousness of the system be determined? How can it be ensured that actors are not acting for their own benefit at the expense of the other and without repercussions, in the absence of a central regulating entity?

This is where blockchain comes in. Its structure is based exactly on a peer-to-peer network, where each network node (a computer, with an individual behind it) communicates directly with all other network nodes. Any transactions that take place between two (or more) nodes are permanently and definitely, thus immutably, recorded in a ledger distributed among all network nodes. Every node in the network keeps for itself an exact copy of the ledger that contains the record of all transactions that have taken place from time zero to the present. This tells us that if there is a malicious actor in the process who decides to alter the ledger to his or her own advantage, all other nodes on the network will be able to see the cheating since the malicious actor's ledger will be different from all the others. Later in this work, how the security of the network is achieved will be explained.

Blockchain allows a trustless environment for disintermediation, lower transaction costs, quick and automated operations, instant settlement of transactions between parties, immutable records registered on a timeline basis that allows the tracking of all past transactions, with full transparency of data without any possibility of failure or fraud, whether intentional or not (Dorri et al., 2017) . In short, a bulletproof governance (UNESCO, n.d.) system of a distributed stakeholder network, unknown to each other.

The main objective of this chapter is to determine whether, in the specific case of carsharing, the market should adopt the so-called sustainable innovation - technologies that enhance attributes that customers and companies already value, such as Web2 developments - or whether it should adopt a disruptive innovation (Bower and Christensen, 1995) - technologies that have different or new attributes that deviate from those that customers have valued in the past, and capable of creating completely new business models or relationships, as is the case with Web3 technologies, specifically blockchain.

For this purpose, a basis for comparison between three proposed solutions for carsharing will be presented, two of them based on a centralized system (B2C and P2P); and another based on blockchain technology where there is complete disintermediation in the business between the parties. Both technologies have their strengths and weaknesses, but what is really important to assess is whether disruptive innovation actually brings new opportunities or whether they are not so relevant and therefore sustainable innovation should be adopted.

In the end, a brief analysis will be presented, (since a detailed one would require a more extensive scientific format), on the challenges of disruptive innovation namely at the level of parallel but necessary technologies, such as the Internet of Things.

2. Theoretical Framework

2.1 Indicators and Market Research

On average, a car spends around 95% of its lifetime parked (Shoup, 2021). This means that the return of usefulness to the car owner is less than 5%, not to mention fixed costs such as insurance, taxes and maintenance.

Passenger cars are a significant polluter, accounting for almost 60,7% of total CO_2 emissions from road transportation in Europe (EU, 2019). A single car can increase CO_2 emissions between 1000 and 5300 kg/year. A more efficient car, like an electric vehicle (EV), will also produce pollutant gas emissions, but in a quantity 3 times smaller than a gasoline or diesel car. It is possible to reduce CO_2 emissions up to 1190 kg/year. Portugal is among the EU Countries with the highest passenger car CO_2 performance - 99 gCo_2/Km - placed in 3rd position only after Norway (47 gCo_2/Km) and France (98 g CO_2 /Km) and way before Poland (123 g/Km). Germany (119 g/KM) (Mathieu and Poliscanova, 2020).

Europe is aligned in its commitment to reduce its carbon footprint by conducting its efforts from the European Green Deal (EC, 2021), so electric mobility will have a strong impact on the desired goals. The impact of carsharing in reducing its ecological footprint has significant effects (Nijland et al., 2015). Adopting this measure can save up to 175 to 265 kg of CO_2/person/year. This is between 8% and 13% of the CO_2 emissions related to car ownership and car use. About half of this reduction can be ascribed to less car use; the other half to the lower degree of car ownership. Every car sharing results in the effect of removing 13 cars from city traffic or deterring the purchase of a new one (Cohen and Muñoz, 2016). Scaling up carsharing, as is the Energy Agreement's[1] intention, will contribute to a reduction in greenhouse gases (Jung and Koo, 2018).

Many European and non-European cities, such as Paris, Barcelona, Berlin, Milan, Seoul are already implementing EV car-sharing programs, as seen from the Annual Report of the International Energy Agency (IEA, 2013).

Emerging applications and services will account for US$203 billion in revenues. The Passenger Economy represents a US$7 trillion global opportunity in 2050 (Lanctot, 2017). It is easy to predict that disruptive technologies as a driving force for innovation will therefore play an important role in the design of new solutions.

---

[1] The Energy Agreement contains a target for 2020, for the transport sector, to reduce CO_2 emissions by 1.3 to 1.7 Mt. Here, only the tank-to-wheel emissions are included. Emissions related to fuel production or vehicle manufacturing and demolition are not included.

2.2 Blockchain Technology

There are many ways to explain what blockchain technology is. But for those who have never heard of this technology, the best way to begin to understand it is that it allows for the removal of trust between parties. That is, one doesn't need to trust any other party, nor rely on a third party of confidence, to know that a particular contract or parameter was conducted from start to finish the way it should have been conducted, without any possibility of failure or fraud. This is exactly what the first blockchain application - Bitcoin - has achieved (Nakamoto, 2008). The ability to transfer value over the internet directly from person A to person B without having to go through a trusted authority (e.g., a bank) that verifies and validates that transaction, that ensures that the value has not been double spent and that it has indeed passed into the hands of person B, and not C. In short, a "bulletproof" governance system of a distributed stakeholder network.

The second, and perhaps most important application of blockchain is 'smart contracts', introduced by the Ethereum network (Buterin, 2014). A smart contract, like any other contract, lays out the terms of an agreement, but unlike a traditional contract, the business logic is written in (computer) code and is self-executable and self-enforceable when the terms of the contract are met. In essence, an if-then function. For example, a vending machine tells people that if one wants a bottle of juice, its price is X, and to get it, they have to deposit X. One makes the deposit, and immediately, gets the juice. If one puts in X+Y, the machine gives back the change Y to the person. There was no party brokering the transaction between the person and the machine (the trust issue), and the person knows in advance that if he holds up his end of the bargain, the machine holds up hers. As simple as that. Many other web3 applications (more precisely, Dapps – Decentralized Applications) have come up since 2008 (Bitcoin) and 2015 (Smart Contracts – Ethereum Blockchain). Cryptocurrencies, protocol tokens, utility tokens, purpose-driven tokens, DAO (Decentralized Autonomous Organizations), NFT (non-fungible tokens), DeFi (Decentralized Finance), Dex (Decentralized Exchanges – P2P Cryptocurrencies trading).

Looking deeper into what blockchain is, how it works, and what it enables, for the sake of a better understanding, a comparison with traditional databases can be made. One important point is that blockchain is not a database, nor does it intend to be. A traditional database works on the basis of a CRUD system - Create, Read, Update and Delete. In comparison, a blockchain can be referred to as an append-only data structure, i.e., it only allows to Create and Read, but never Update or Delete any given data, past or present. Also, databases are centralized.

Even those who are distributed among external servers (e.g., Facebook) are anyway centralized because they rely on a central authority (Meta). This creates a single point of failure in the system.

In opposition, a decentralized ledger - a distributed ledger - can be more decentralized, less decentralized, or fully decentralized; is distributed across all participants in the blockchain network, and where each of them owns an exact copy of the ledger. Thus, if participant A decides to "say" he has received $100 from participant B, when in fact he received $10 (trying to profit $90), all other participants in the network will know that participant A is trying to commit fraud, because his ledger ($100 from B to A) is different from all the others ($10 from B to A). The fundamental principle is that the majority of the participants in the network reach a consensus on the true ledger. But even then, how can one know which is the true ledger? Perhaps the majority of the network is trying to fool the minority. Or vice versa. It is not impossible. This is prevented by the so-called *consensus mechanism*, purely based on mathematics and cryptography.

A blockchain ledger can be defined as a set of time-sequenced, interconnected blocks (block-chain), where each block contains the transactions that are performed on the network. The procedure of chaining the blocks together is called chaining (hence the term block-chain), where all data in a block is run through a special function called a cryptographic hash. Cryptographic hashes create a unique output for a specific input. Therefore, the hash of each block will always be unique based on the inputs. To link blocks together, the header of the current block contains the hash of the last validated block. Any attempt to change the data of any past blocks in the blockchain will result in a completely different hash and the new hash will not match the hash in the next block header, thus breaking the chain and invalidating all blocks linked to where the change was made. This is what gives blockchain the property of immutability – data can't be changed – making it highly fraud-resistant. This means that, securely and reliably, one can query and track all transactions that have occurred within the blockchain, from time zero to the present, with the assurance that nothing has been modified. It is difficult to find this security and reliability in a traditional database.

However, the security and reliability of transactions within the blockchain network is not limited to this. It is mandatory to prove beyond doubt that the transaction came from one place to another. This is achieved through so-called public key cryptography, where there are two key pairs, one public and one private. The public key is derived from the private key via a cryptographic hash function, but the private key can never be derived from the public key. Transactions are sent from public key to public key (actually from address to address, where each address is derived from the respective public key, again via a hash function), and where a digital signature is used to prove that the public key matches the private key. The digital signature is used to sign/approve any transaction carried out by the holder of the public/private key pair. All transactions submitted on the blockchain are signed using the private key and verified using the public key. In short, with complete confidence and certainty, the recipient of the transaction knows that the sender is who he says he is and actually has the transaction funds. The recipient will receive the funds and - as long as he does not disclose his private key to anyone else - can access those funds.

The way to prove that the funds of a transaction (a) actually exist, and their provenance, (b) that the recipient actually receives them in their entirety when he attempts to access them, and (c) that this verification and validation is achieved not only between the sender and recipient, but by all participants in the network, is through a *consensus mechanism* (everyone gauges the single source of truth) called *proof-of-work,* which is based on the security of cryptography, the randomness of large numbers, computational calculation and motivational game theory. There are other consensus mechanisms, such as *proof-of-stake,*[2] but the principle is the same.

In essence, proof-of-work makes the economic cost of attacking or defrauding the system disproportionate to the benefit of doing so. The mechanism assumes that all nodes in the network can potentially be corrupt, and that the lowest common denominator is money. The proof-of-work is designed so that if one applies money and plays by the rules, he/she can earn network tokens, and that it does not pay to cheat the system, since mining requires specially designed hardware for brute-force computational calculations and consumes large amounts of power. When tokens are sent across the network, each node in the network can propose new entries to be added to the ledger. These nodes validate the transactions and compete with each other to solve a complex computational puzzle. In this process, they have to collect all recent transactions from the network, including some additional metadata, verify the transactions, guess a pseudo-random number (*nonce*), and run all the data through a cryptographic algorithm (SHA-256) to find the hash of the new block.

[2] Ethereum uses proof-of-stake, where validators explicitly stake capital in the form of ETH into a smart contract on Ethereum. This staked ETH then acts as collateral that can be destroyed if the validator behaves dishonestly or lazily. The validator is then responsible for checking that new blocks propagated over the network are valid and occasionally creating and propagating new blocks themselves.

This means that they have to perform computational work, which is why this process is referred to as "Proof of Work". If a particular node in the network is the first to find this hash value, it can add the block to its ledger and transmit to the rest of the network the hash value of the new block, including all the data in the block. The other nodes can now check the validity of the hash, adding the new block to their copy of the ledger. The proof-of-work is designed so that even though the hash is difficult to find, the solution found is easily verified as true. By participating in this competition of finding the hash value first, mining nodes collectively ensure that all transactions included in a block are valid. The winning node is rewarded with the "block reward" in the form of newly created network tokens. This is the reason why the process is referred to as "mining". The hash of a validated block therefore represents the work done by the miner.

If a malicious miner were eventually the fastest computer to find the hash (an unlikely situation, but theoretically possible), the rest of the network would not accept its block of transactions. The cheating miner would therefore not receive the block reward, even if it invested enough computational power and energy to do so. This is an economic measure to deter attacks on the network. And even if the attack on the network should be successful, it would become prohibitively expensive in terms of computing power, energy consumption and time, and most possible the gains would not overcome the losses.

Much more can be said to explain blockchain or its different applications, but what has been described below summarizes how blockchain works and what it guarantees:

- Immutability of data
- Publicly verifiable (Accountability to costumers and end-users)
- Quality assurance (e.g., Supply chain auditing and tracking)
- Low transaction costs (removing middleman reduces costs)
- Tokenization (the digital twin of a physical asset, traded over the internet)
- Redundancy and high fault-tolerance
- No centralized authority
- Low barrier to entry
- Instant, global transactional settlement
- No double spending

However, as any other technology, the benefits it provides come at a cost. There are some drawbacks to blockchain that must properly considered in order to determine if blockchain is a good choice in an overall solution architecture. Decentralization, security, redundancy and high fault-tolerance, although seen as potentially preferable, make blockchain extremely inefficient (for the time being) when compared to legacy IT systems, which usually facilitate speed and scalability, but at the cost of being less secure. It all comes down to the so-called scalability trilemma, the trade-off between security, decentralization and scalability itself. Decentralization is the premise of a distributed network, security is the most important aspect when the network involves a set of untrusted actors, and scalability refers to the number of transactions a system can process per second. Today, scalability of public blockchain networks is one of the main bottlenecks regarding mass adoption and also one of the most worked on R&D issues. The implementation of a blockchain system will always depend on the combination of these three factors, and their relative importance to the architecture of the solution.

Another key aspect when considering blockchain implementation is what type of blockchain should be used. A good way to do that is to determine whether all participants are considered equal or whether there is a need to differentiate according to who has the right to read (open vs closed blockchains) or write (public vs private blockchains) in the ledger and who is permitted to do so (permissionless vs permissioned blockchains).

It is now easier to understand the scalability trilemma. A private blockchain, where the network actors are known up front, is itself more centralized but less secure. But once the actors are considered to be trustworthy, there is no harm in giving up some security if we get more performance for it. On the other hand, a public, decentralized blockchain, where the actors are unknown and therefore seen as untrustworthy, will have to have a mechanism (proof-of-work) that gives it maximum security, but most probably having to give up scalability and throughput for this purpose. All choices depend on the specific case, business logic and scope.

The second major application of blockchain is the so-called smart contracts, self-executing agreements, materialized in the form of software. The code contains a set of rules according to which the parties to that smart contract agree to interact with each other. At any point in time when the predefined rules are met, the agreement is automatically executed by majority consensus of the blockchain network. Smart contracts provide mechanisms for the efficient management of tokenized assets and access rights between two or more parties. One can think of this as a cryptographic vault that unlocks value or access if and when specific predefined conditions are met. Smart contracts therefore provide a public and verifiable way to embed governance rules and business logic into a few lines of code, which can be audited and enforced by the majority consensus of a P2P network.

Smart contracts provide:

- Autonomy: Smart contracts can be developed by anyone, with no need for intermediaries such as lawyers, brokers or auditors.
- Backup: A blockchain and smart contracts deployed by it can provide a permanent record, allowing for auditing, insight and traceability even if the creator is no longer in business.
- Efficiency: Removing process intermediaries often results in significant process efficiency gains.
- Accuracy: Replacing human intermediaries with self-executable code ensures the process will always be followed accurately.
- Cost savings: Removing intermediaries often provides significant cost reduction.

Smart contracts can provide greater transaction security than traditional contract law, thereby reducing the costs of coordinating the audit and enforcement of such agreements. Smart contracts also circumvent the "principal-agent" dilemma of organizations, providing more transparency and accountability, and reducing bureaucracy.

Among many other things, smart contracts can be used to create and manage cryptographic tokens that can represent any asset or access right and even incentivize behavior. Tokens could emerge as one of the most important applications of smart contracts, potentially revolutionizing asset management as we know it.

Smart contracts and distributed ledgers could also be a catalyst for machine-to-machine settlement in the context of the "Internet of Things." However, this requires that all objects in such an "Internet of Things" have a blockchain identity and can thus be addressed uniquely. The addressability of each machine or other physical object needs to be tamper-proof. This can be achieved by tagging or registering objects with a so-called "crypto accelerator," which is also referred to as a "digital twin." A crypto accelerator is a small micro-controller optimized to run the most important cryptographic algorithms. It can be the size of a sticker on a piece of fruit and thus serve as the basis for use cases such as supply chain transparency. With a digital twin, any physical object can send unique digital signatures, or send and receive tokens. With the current pace of development of this technology along with the convergence of other emerging technologies such as IoT, Big Data, and Artificial Intelligence, it can be predicted that in the future individuals, organizations, and machines can freely interact with each other with little friction and at a fraction of today's costs (Voshmgir, 2020).

2.3 *Blockchain for Carsharing – The Motivation*

IT and online platforms facilitated the rise of the sharing economy, offering motivation for consumer participation such as economic benefits – cost and time savings -, convenience and social value. That is the premise of carsharing. Individuals able to benefit from private car use without the costs and responsibilities of ownership, while society potentially benefits from more efficient vehicle usage.

The European shared-mobility market has gained popularity over the past few years and is projected to grow from an estimated market size of € 70 billion in 2022 to € 150–200 billion by 2030 (McKinsey, 2022). Reducing carbon emissions, vehicle trips and traffic congestion, ridesharing allows getting rid of vehicle ownership and related costs, such as maintenance and depreciation, making transportation more reliable, convenient, enjoyable and safe.

But carsharing[3] only accounts for around 3 billion of total shared-mobility revenues, and the way it is currently conducted has caused many projects to fail. Among the main reasons are the mandatory fixed pickup and drop-off points, parking fees included on the rent (sometimes higher than what is used), and tank refilling. Specifically regarding the P2P model, trust and confidence become the main pain points. All of these issues can be tackled with blockchain technology, as it will be seen later, especially when compared to traditional car sharing systems.

Given its incipiency, it is not yet clear whether blockchain technology will be the driving force to leverage these values and respective degree of carsharing usage, but what is known is that disruptive innovations offer new attributes or technological features, even when these features are seemingly undervalued or are initially rejected by users. Thirty years ago no one would have guessed what the Internet would be able to provide; what is certain today is that in the course of the past 30 years, the Internet and its applications have radically changed business and people's lives. The internet was part of was Web1 (the internet of information) and Web2 (the internet of social and e-commerce platforms) technologies. We are now in the phase of figuring out to what web3 (the internet of value transfer) and blockchain-based applications will be able to offer. The odds look favorable. The goal of this chapter is to provide an insight into how blockchain can add value to the P2P model of carsharing when compared to legacy technologies currently, and broadly, in use.

[3] Carsharing services may follow free-floating, station-based or peer-to-peer models. Shared mobility includes taxi/ride hailing (like Uber or Lyft), pooled shuttle, robot-shuttle, carsharing, car rental, and shared micro mobility.

3. Research Methodology

P2P car sharing is gaining ground, but still has some bottlenecks; multiple parties are involved in the process, so there is a need for a third party to establish trust and handle all the complex work of billing and revenue sharing between those parties. As already outlined, blockchain technology allows these functions to be conducted directly, and therefore more quickly, without the need for a costly intermediary. Current C2C models also include other inefficiencies relating to insurance (the renter has to purchase an insurance policy for the trip at a significant premium from the C2C operator), transferring car keys, arranging a time to meet a user, mandatory access and drop-off points or the obligation to deliver the vehicle with a full tank, just to name a few.

Here a solution for a decentralized P2P carsharing platform will be proposed, and its advantages and potential assessed, as compared with the current centralized C2C platforms. To do this, it is first necessary to clarify how identities and the "coin" of exchange occur within the blockchain. The proposed solution will be named PAMS – Personal Autonomous Mobility Solution.

3.1 Blockchain Wallets & Decentralized Identities

Users can create and register a Decentralized Identification (DID) when activating a blockchain wallet, which creates a pair of private and public keys. Public-key cryptography is used for authentication and encryption. Only the private key can prove one's identity. The private key acts as your personal lock on the wallet.

Any DID can be linked to attestations (verifiable credentials) that are issued by other people and institutions attesting specific characteristics for an identity owner, such as name, address, email, age, or driver's license. The credentials are signed by their issuers using public-key cryptography. Once signed by an issuer, credentials can be managed using the wallet of the identity owner directly.

The separation of the "identifier", "authentication", and "data" is crucial to a user-centric setup. It can be seen as a system of checks and balances in a data-driven economy that guarantees the level of autonomy and privacy over one's digital footprint.

The wallet acts as a personal container that allows the control of digital identities. The wallet is the digital equivalent to a physical wallet, which usually acts as a container for all the ID cards, such as driver's license, bank card, national ID card, social security card, in addition to money.

Just as anyone opens their wallet to reveal an ID card, the Web3 wallet needs to be activated to reveal someone's digital credentials to third parties.

No one can see the contents of the Web3 wallet without user consent. The content of the wallet remains concealed until you choose to reveal something to a third party. The digital wallet is portable, as a dedicated hardware device or an app in someone's mobile phone or notebook.

3.2 The Token Economy - How Transactions Occur

In a blockchain system, transactions cannot be conducted through fiduciary money (Fiat). Instead, a digital money system is used, called a token. A token can be fungible (can replace another means of payment), or non-fungible (representing an asset with unique characteristics, such as a vehicle or other types or properties). out of the scope of this chapter.

As already seen, a smart contract is a self-execution agreement, materialized in the form of software. The code contains a set of rules according to which the parties to the smart contract agree to interact with each other. The communication is made between the parties' wallets. It provides mechanisms for efficient management of tokenized assets and access rights between two or more parties, ensuring:

- **Observability** – The participants of the smart contract should be able to see how well their counterparts abide by the terms.
- **Enforceability** – Ensure that the agreed terms are being fulfilled by using various measures.
- **Verifiability** - In case a conflict arises between the participants, the contract should be auditable.
- **Privacy** - Data of the smart contract should only be available to the participants of the smart contract, otherwise encrypted.

With regard to carsharing, the smart contract grants access to the EV upon a given condition (for example, locking up funds for payment, or driver's license), all auto-executed without intervention from both parties, EV owner and the driver. When the driver decides to end his trip, the smart contract will debit his wallet for the time and distance travelled, as well for the electricity he has consumed. If any of the conditions are not met, the contract will not be executed.

The "EVM" Non-Fungible Token: Each vehicle listed on the marketplace is represented with a non-fungible token (NFT). It is unique in nature, allowing the inclusion of metadata for describing distinguishing properties such as vehicle ownership, brand, model, year, license plate, inspection certificate, etc. Whenever a new vehicle is added to the marketplace a new token will be minted and transferred to its owner. The token will contain identifying information in its metadata, linking it with the vehicle it represents. At any given time, the holder of an EVM token will be granted sole access to the corresponding vehicle and, as such, the ownership indicates that a rental transaction has been entered into. The token is returned to the vehicle owner on completion of a rental transaction.

The "PAM" Stable Token: Given the market volatility of fungible tokens listed on centralized exchanges (Reiff, 2021), an in-app stable token - which we will call it PAM - can be adopted. Each token will be divisible by 18 decimal places for conducting micro-transactions and its value is pegged to Nation-State currency on a 1:1 parity. Any value exchange on the PAMS marketplace - renting an EV, charging the EV at a charging dock, etc., will be conducted using the PAM token. PAMS platform will also provide a Fiat-PAM gateway to facilitate the easy exchange of GBP, EUR, USD and other national currencies, especially for those who have no knowledge about cryptocurrencies or how to buy them.

3.3 Managing Identities

Decentralized identifiers enable users and devices to authenticate themselves and manage values exchanged between other network participants. Each PAMS entity has its own identity. These entities can be (Figure 1);

- EV Drivers
- EV Owners
- The EVs

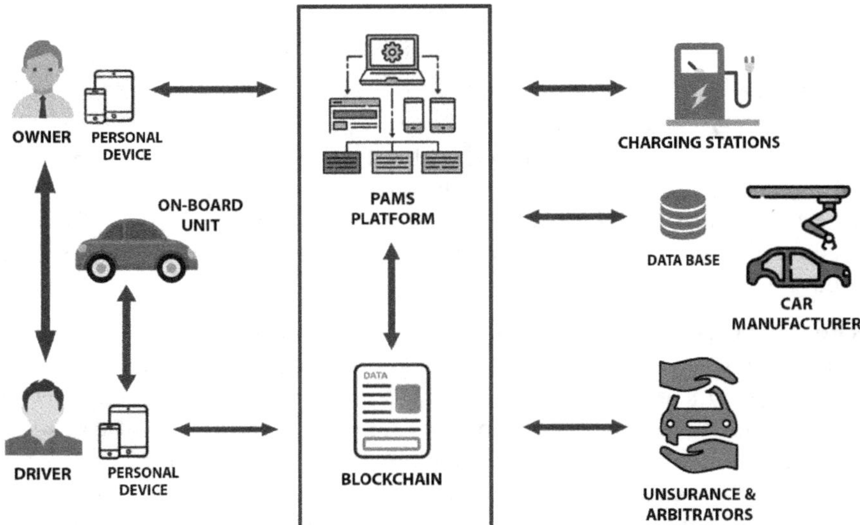

Figure 1. Overview of the identities involved in a decentralized P2P Carsharing solution.

- Charging Docks
- Insurance Companies
- Arbitrators

4. PAMS Smart Contract - Carsharing Contract

1. The owner of the car deposits his EVM token in the smart contract, in order to list his car on the platform. EVM will contain metadata such as car model, year, license plate, and other pertinent data (Figure 2). Car images will be stored off-chain in IPFS (IPFS, 2022a), which is decentralized storage, to reduce on-chain transaction costs. Storing IPFS URI (IPFS, 2022b) is much more effective and cheaper than storing images on the blockchain network.

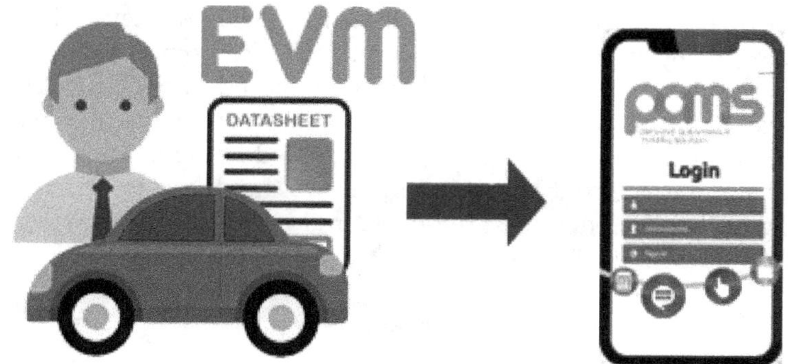

Figure 2. Submitting vehicle and owner identity to the blockchain.

Blockchain vs Legacy

Whatever the system used; all the vehicle data is available on a front-end platform to be browsed by those who want to rent the EV. But blockchain technology brings the convenience to the renter of being able to upload all the relevant information without having to go through a manual process of submitting the data to a centralized third party. No paperwork as everything is digitized and rules are coded. However, there is also the need to use oracles to verify the veracity of the information with agencies such as the DMV or the vehicle registration office.

2. After choosing the vehicle he wants to rent, its price, and determining its position via GPS, the driver (person who wants to rent the car) sends a request to rent the car by locking up - but not effectively paying - a given amount of PAM tokens (using a FIAT-PAM gateway) in the deployed smart contract for insuring the payment at the end of the contract (Figure 3).

Figure 3. Locking up funds to the smart contract in order to initiate the rent.

Blockchain vs Legacy

There are no upfront expenses or membership costs as there usually are in traditional systems, be they B2C or C2C. Also, the owner of the vehicle can determine at will the price he wants for the rental, without having to meet a price imposed by a third party, or commission fees.

3. The owner signs the booking details using his private key and encrypts them using the driver's public key. Only the owner and the driver have access to the sensitive information stored in booking details, as these details are discussed offline and only stored in the smart contract when encrypted (Figure 4). Once encrypted he sends these details to the storage of smart contract and allows the driver to access the car (Figure 5). The encrypted details stored in the smart contract can be accessed by the driver and decrypted using his private key.

Figure 4. Secure key transference.

Figure 5. Vehicle access.

Blockchain vs Legacy

Unlike legacy systems, in which the vehicle keys have to be handed over, with blockchain the access to the car is done automatically and in a totally secure way, without the need for scheduling a time for the vehicle to be accessed. Currently there are telematics-based solutions that allow access to the vehicle without requiring physical keys, but the problem lies in security, since these systems can be hacked. With blockchain, it is virtually impossible for this to happen.

4. The driver accesses the car and uses it for the amount of time and distance he wants (Figure 6). When the trip is over, the driver ends the car rental process by dropping off the owner's car at a previously determined boundary (e.g., within city limits), but with no specific drop-off point.

Figure 6. Pay per exact distance, time, electricity and parking.

Blockchain vs Legacy

Unlike legacy systems, where you usually pay by the day or even by the hour, with blockchain you only pay for the distance you traveled and the time you used the vehicle. No more, no less. This data, recorded in the blockchain, becomes immutable and traceable, so both the vehicle owner and the driver will contribute to, and be assured of, the reliability and certainty of this data. However, there is a need to use IoT to establish communication and data upload between the vehicle and the blockchain.

5. Driver's wallet will be debited for the time, distance and electric energy consumed. Should the driver need more charge, he can supply the EV at a charging station, and the amount that was not used during the trip will later be credited. to his wallet.

Blockchain vs Legacy

Here again there is a need to use IoT technology, but it eliminates one of the many bottlenecks of carsharing, which is the obligation to return the vehicle with a full tank. And once again, the owner of the vehicle will know, through blockchain, that the actual amount of electricity was consumed.

6. Once the process is over, both the driver and the owner can separately withdraw their deposits and earnings (Figure 7).

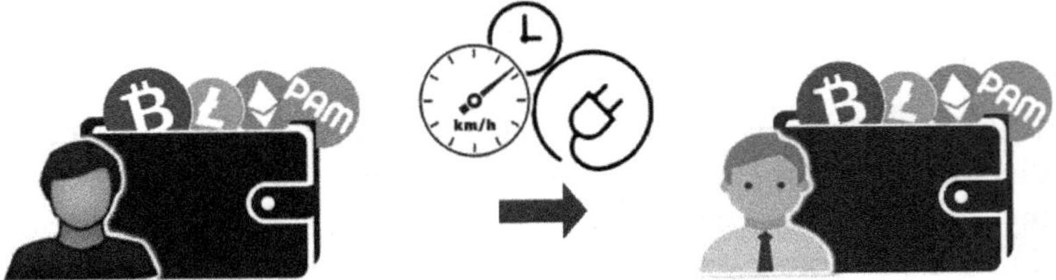

Figure 7. Instant payment settlement.

Blockchain vs Legacy

Blockchain provides self-enforced instant payment settlement between parties. That is not possible in traditional systems.

7. During his trip, in the event a car accident or damage occurs, identified through IoT technology and recorded permanently on the blockchain, a separate smart contract between him and the insurance company is deployed.

Blockchain vs Legacy

Smart contracts are auditable by nature, and in case of an incident involving the owner or the driver, the smart contract can be audited to reveal information about them. And since the information is traceable, it will be known immediately who was driving the car at the time of the incident. Without the use of these technologies, a minor incident that does not even involve the police would not be immediately recorded, and it would be more difficult (in some cases, even impossible) to identify who was driving the vehicle.

5. Results

A basis for comparison between legacy systems and a blockchain-based system can be drawn. Since the aim is to assess whether blockchain will increase the adoption of carsharing, which always involves factors of uncertainty and mistrust, the comparison will be made around three groups that have a direct influence on the decision: convenience, cost efficiency, and security.

	Blockchain	Legacy	
	P2P	C2C	B2C
CONVENIENCE			
EV can be owned and shared by individuals, whenever and wherever they want. Freedom of choosing whether or not to rent, and when	Yes	No	No
No mandatory access or drop-off points	Yes	Not entirely	No
No prior booking scheduling	Yes	No	No
No need for car keys delivery	Yes	Not entirely	No
No need for tank refilling	Yes	No	No
No paperwork fill out	Yes	Yes	No
COST EFFICIENCY			
Pay only and exactly for what you use (distance, time, electricity)	Yes	No	No
No intermediation costs	Yes	No	No
No upfront payments or membership costs	Yes	No	No
Instant payment settlement between parties	Yes	No	No
No price definition by third party	Yes	No	No
Lower premiums through usage-based insurance payment and reduced processing overheads in claims handling	Yes	No	No
SECURITY			
Car keys security against hacks	Yes	No	No
Real time auditing and data transparency by all parties	Yes	No	No
Immutable and irreversible data	Yes	No	No
Data transparency and traceability	Yes	No	No

6. Conclusions, Limitations and Future Directions

The social, environment and financial benefits of P2P carsharing align with the benefits of the sharing economy: increased use of existing resources and more human connection. P2P carsharing represents an evolution in the sharing economy in which shared mobility has transitioned from the sharing of a commercial vehicle fleet to the sharing of personal vehicles.

As demonstrated, blockchain technology brings significant improvements by providing more convenience, cost efficiency, and security for both parties, which will certainly lead to greater adoption of carsharing, and thus to cities with less traffic, fewer carbon emissions, and an overall improved quality of life. Also, new business models or opportunities can arise, for example, drivers can earn credit tokens for giving out their driving experience, if they will to do so. That will also work as an incentive to adopt carsharing. An insurance company registered on the network can pay directly to the driver for valuable data over driver's behavior; electricity providers can give away discounts if drivers charge the EV at their charging stations; municipalities can be paid by the minute for the time an EV is parked, providing fair payment to both municipalities and drivers. Besides carsharing, blockchain can enable further market segments worth paying attention like driverless cars. Another solution could be a decentralized energy generation and peer-to-peer energy trading where EV could be charged and help transport energy across micro-grids.

There are however some limitations that have to be addressed, although not impossible to overcome. One of them is the issue of paid parking, specifically when leaving the vehicle. One solution might be for municipalities not to require payment, since carsharing is potentially of interest to them as well. Another could be to force the driver to park somewhere unpaid when he leaves the vehicle, and if that doesn't happen, a new smart contract is triggered automatically charging the driver for the parking time (funds must be locked until a new driver comes in). Of course, this requires the creation of adequate IoT infrastructures, which can be a clear impediment to the advantage of being able to leave the car wherever one wants.

Another limitation, as already mentioned, involves insurance companies, which will have to design products tailored to decentralized carsharing. But as a general rule, like banks, insurance companies are already studying use cases for blockchain technology, so in the future it is expected that the insurance issue will not be a limitation, but rather a new market opportunity.

The use of blockchain technology itself is still incipient and highly costly largely due to the lack of human resources with the necessary skills or even legal regulation, but like any disruptive technology, the market will eventually self-adjust. Add to that the fact that it requires parallel technologies such as Internet-of-Things, Big Data or Artificial Intelligence.

Nevertheless, disruptive technologies are often redefining how companies conduct business or how products are offered or used.

Blockchain can potentially help create more value, achieve operational efficiencies, and improve the customer experience.

References

Akande, A., Cabral, P. and Casteleyn, S. (2020). Understanding the sharing economy and its implication on sustainability in smart cities. Journal of Cleaner Production, 277: 124077.

Belk, R. (2010). Sharing. Journal of Consumer Research, 36(5): 715–734.

Bernardi, M. and Diamantini, D. (2018). Shaping the sharing city: An exploratory study on Seoul and Milan. Journal of Cleaner Production, 203: 30–42.

Botsman, R. and Rogers, R. (2010). What's mine is yours. The Rise of Collaborative Consumption, 1.

Bower, J.L. and Christensen, C.M. (1995). Disruptive Technologies: Catching the Wave.

Buterin, V. (2014). A next-generation smart contract and decentralized application platform. White Paper, 3(37): 2–1.

Cohen, B. and Munoz, P. (2016). Sharing cities and sustainable consumption and production: Towards an integrated framework. Journal of Cleaner Production, 134: 87–97.

Dorri, A., Steger, M., Kanhere, S.S. and Jurdak, R. (2017). Blockchain: A distributed solution to automotive security and privacy. IEEE Communications Magazine, 55(12): 119–125.

European Commission (EC). (2021). A European Green Deal. https://ec.europa.eu/info/strategy/priorities-2019-2024/european-green-deal_en.

European Parliament (EU). (2019). Co2 emissions from cars: facts and figures. EU News. https://www.europarl.europa.eu/news/en/headlines/society/20190313STO31218/co2-emissions-from-cars-facts-and-figures-infographics.

Hahn, R., Ostertag, F., Lehr, A., Büttgen, M. and Benoit, S. (2020). I like it, but I don't use it: Impact of carsharing business models on usage intentions in the sharing economy. Business Strategy and the Environment, 29(3): 1404–1418.

International Energy Agency (IEA) (2013). Global ev outlook: Understanding the electric vehicle landscape to 2020. https://www.iea.org/.

IPFS. (2022a). https://www.ipfs.io.

IPFS. (2022b). Best Practices for Storing NFT Data using IPFS. https://docs.ipfs.tech/how-to/best-practices-for-nft-data/#types-of-ipfs-links-and-when-to-use-them.

Jung, J. and Koo, Y. (2018). Analyzing the effects of car sharing services on the reduction of greenhouse gas (GHG) emissions. Sustainability, 10(2): 539.

Lanctot, R. (2017, June). Accelerating the Future: The Economic Impact of the Emerging Passenger Economy. Strategy Analytics. https://newsroom.intel.com/newsroom/wp-content/uploads/sites/11/2017/05/passenger-economy.pdf.

Mathieu, L. and Poliscanova, J. (2020, October). Mission (almost) accomplished. Carmakers' race to meet the 2020/21 Co_2 targets and the EU electric cars market. Transport & Environment. https://www.transportenvironment.org/wp-content/uploads/2021/05/2020_10_TE_Car_ CO_2 _report_final-1.pdf.

McKinsey. (2022, July 5). Snapshot of the European car-sharing market. https://www.mckinsey.com/features/mckinsey-center-for-future-mobility/mckinsey-on-urban-mobility/snapshot-of-the-european-car-sharing-market.

Münzel, K., Piscicelli, L., Boon, W. and Frenken, K. (2019). Different business models–different users? Uncovering the motives and characteristics of business-to-consumer and peer-to-peer carsharing adopters in The Netherlands. Transportation Research Part D: Transport and Environment, 73: 276–306.

Nakamoto, S. (2008). Bitcoin: A peer-to-peer electronic cash system. Decentralized Business Review, 21260.

Nijland, H., Van Meerkerk, J. and Hoen, A. (2015). Impact of car sharing on mobility and CO_2 emissions. PBL Netherlands Environmental Assessment Agency. PBL Publication, (1842).

Reiff, N. (2021, August). What are centralized Cryptocurrency Exchanges? Investopedia. https://www.investopedia.com/tech/what-are-centralized-cryptocurrency-exchanges/.

Shoup, D.C. (2021). The High Cost of Free Parking. Routledge.

UNESCO. (n.d.). Concept of Governance. http://www.ibe.unesco.org/en/geqaf/technical-notes/concept-governance.

Voshmgir, S. (2020). Token Economy: How Blockchains and Smart Contracts Revolutionize the Economy, 2nd. BlockchainHub Berlin.

CHAPTER 11

A Proposal for a Blockchain-Based Electronic Election System

*António Manso** and *Célio Gonçalo Marques*

1. Introduction

The Internet has allowed the development of various technologies that have enabled the digitalisation of services provided by governments, such as requests for clarifications and information, or the payment of taxes and fees, where citizens now consume these services through portals (Maesa and Mori, 2020). These services improve the efficiency of other services and make them easier to use. The electoral system is one of the services that, due to its importance for democracy, raises a set of important trust and security challenges. Cryptography solves security issues that include confidentiality, integrity and authenticity. This technology is widely used in secure computing systems, and has proven to fulfill its goals. Trust can be achieved through blockchain technology that has been showing its potential in several areas of application, the best known of which are virtual currencies.

The digitalisation of voting allows for considerably more economical systems as it eliminates the logistics required for manual voting: Ballot boxes, paper ballots and the staff needed to ensure regular operation (Russell and Zamfir, 2018). For the voter, electronic voting allows remote

LIED, Polytechnic Institute of Tomar, Portugal.
Email: celiomarques@ipt.pt
* Corresponding author: manso@ipt.pt

voting from his/her equipment, avoiding the need to go to the polling station. This feature is particularly important for voters with reduced mobility or voters with difficulties in accessing polling stations, such as remote areas of the territory or those outside the country.

Electronic voting can be an advantageous step towards the digitalisation of one of the most important processes in a democracy. However, it must be ensured that the whole process takes place safely and without interference from third parties. Therefore, our research question is: How to create a secure electronic voting system based on blockchain technology?

Under the scope of the curricular unit of Distributed Systems of the Degree in Informatics Engineering of the Polytechnic Institute of Tomar this question was presented to students. It is usual practice to set challenges based on problems they identify in real life. All these features were discussed and it was drafted how they could be implemented. In groups, students started programming the outlined system for further testing. Several weaknesses have been identified in the proposals submitted, despite all the work done. As a way to overcome the challenge with the collaboration of the students, we started the development of a new proposal in response to the defined requirements, reaching a solution that is characterised by 3 components: Electoral Comission, Electors and Ballot Box. This paper presents a proof of concept on blockchain-based electronic voting system as designed in the Polytechnic Institute of Tomar (Portugal).

2. Secure Electronic Voting System

Blockchain technology, with its first practical application being presented by Satashi Nakamoto in the paper "Bitcoin: A peer-to-peer electronic cash system" (Nakamoto, 2008) has proven to be a secure and reliable technology for supporting a virtual currency. This technology allows an immutable set of facts to be maintained in a decentralised manner through a Peer to Peer (P2P) network, whose decisions are made by consensus, allowing untrusted entities to transact with each other securely without the presence of a third party. The development of Distributed Ledger Technology (Natarajan et al., 2017), which allows the registration, synchronisation and sharing of information between nodes of a P2P network in a secure and reliable way, and of smart contracts (Alharby and Van Moorsel, 2017) which allow the execution of programs in an automatic way when the pre-conditions are fulfilled, made possible the development of applications in the most diverse areas, such as, healt care, identity management systems, access control systems, decentralised notary, and supply chain management (Maesa and Mori, 2020).

The security and reliability required for an electronic voting system make this technology very attractive for this type of application, and several proposals have been put forward (Huang et al., 2021). Most of these applications use the services provided by public domain blockchains such as Bitcoin or Ethereum (Maesa and Mori, 2020) for vote storage, and differ in the way the vote is introduced into the system.

There have been experiments applying blockchain technology to electronic voting systems in different countries, such as the US, Russia, Japan and Switzerland (Jafar et al., 2021; Park et al., 2021). In these experiences, situations of vulnerability were identified that put into question the validity of the process. In Portugal, a pilot study was carried out in the district of Évora during the last European elections. However, it was necessary to go to the polling station with voting machines and print a ballot paper to be inserted in the ballot box (Braz, 2021). As stated by Joseph Stalin "Those who vote decide nothing. Those who count the vote decide everything." (Monteiro, 2019, p. 8). This is the main reason why many countries are reluctant to move to electronic systems, as low-tech paper ballots remain the most secure system. Problems may occur, but to a minor scale, and there is also the possibility of auditing and verifying the vote count as often as necessary (Park et al., 2021).

It is important to understand the requirement for an election system. IT is necessary to guarantee (Jafar et al., 2021):

- **Anonymity/Vote Privacy:** Within the system all votes must be anonymous, i.e., it must not allow the identification of the person who submitted the vote.
- **Auditability and Accuracy:** The data must correspond exactly to the votes submitted. It should be impossible to change any vote. It must have the possibility of being audited to verify its correctness.
- **Democracy/Singularity:** In a democracy there are criteria that define who can vote, so only those who are eligible to vote can do so and only once each election. No ballot can be duplicated.
- **Availability and Mobility:** It should be available at all times during voting period, and accessible to all potential voters.

For all these reasons, it is a very complex system to implement electronically. It requires checking the legibility of those who access it, as well as maintaining its anonymity, without losing the possibility of being audited externally (Jafar et al., 2021; Park et al., 2021).

Despite the challenges that electronic voting can present, there are undoubtedly benefits: "E-voting is cheaper, increases voter turnout and confidence, and transforms a process that has existed for thousands of years into one that is exceptionally efficient and on par with cultural and technological advancements" (Monteiro, 2019, p. 7).

3. Methodology

Based on the challenge presented, a proof of concept on blockchain-based electronic election system was proposed to students. The objective of this work was the development and testing of an electronic voting system that can possibly replace the current system of in-person voting.

We adopted an agile methodology for this project, and introduced it in the design of the programmatic contents that were taught throughout the course. The students started by collecting the system requirements that were implemented as new knowledge was acquired. The requirements collection was based on the Portuguese electoral system and the analysis of the electoral data available on the website of the National Electoral Commission (https://www.cne.pt/).

We identified the following features that the system should possess:

- Each election can have several ballot papers that are made public before the election starts.
- Each ballot paper has a set of candidates.
- Only registered voters can vote.
- Each voter can vote only once.
- Voters may choose only one candidate on each ballot paper or choose none by allowing a blank ballot.
- Votes are anonymous.
- Votes are confidential during the period in which the election is taking place.
- The voter can confirm whether their vote counted towards the outcome of the election.

The system should also have a number of features to ensure user confidence and system security.

- Ballots should be certified to prevent tampering and fraud.
- Votes should be certified to ensure that they have not been altered.
- Votes shall be stored in a secure, reliable and auditable system.

Unit tests were used for the validation of the components using automatic test frameworks in each phase of the project implementation, as well as usability tests using fictitious elections for system verification. The final test was carried out with a vote to set the date for the presentation of projects. All students of the curricular unit used the system to choose a date by secret ballot.

4. Proof of Concept Results

This chapter describes the final version of the prototype built through several iterations using agile methodology. It begins by listing the technologies used to ensure compliance with the system requirements, then the system architecture and its implementation are described, and it ends with the presentation of the tests of the system.

4.1 Technologies Applied

To ensure the security, trust and integrity of votes we used blockchain technology. The blockchain architecture allows facts to be stored in a distributed database immutably, where decisions are made by consensus.

Data is stored in blocks that are connected to each other forming a chain of blocks where each block has :

- Previous Hash - Hash code of the previous block that enables the chaining of blocks.
- Facts - Immutable information that is intended to be stored securely.
- Hash - Data verification code given by a cryptographic hash function, such as SHA256, to verify the integrity of facts.
- Proof - A consensus mechanism (Zhang et al., 2020) that validates the previous data, such as proof of work (pow) and proof of stake (pos).

The security of the data of each block is given by hash functions that guarantee that the data was not corrupted and the Proof field the consensus between the nodes of the network about the validity of the block and consequently its security.

Being a distributed database, where each node of the network has a copy of the facts stored in the blockchain, the decisions made by consensus, where at least 50% of the participants agree with the answer and each piece of information has mechanisms to validate its integrity, this technology is resistant to informal attacks and becomes suitable for the storage of votes in an election.

The confidentiality of the vote is achieved through encryption with asymmetric keys. The vote is encrypted with a public key, accessible by everyone, and can be decrypted with a private key. The private key is disclosed at the end of the election, which allows votes to be confidential while the election is in progress, and accessible after the key is published.

The integrity of the vote is achieved through the digital signature of the ballot paper by the relevant authorities. As the vote must be confidential the authority cannot have access to the content, so the blind signature method proposed by Chaum(1983) was used.

4.2 System Architecture

In the implementation of the system, we identified three independent components (Figure 1), which interact with each other during electronic voting. Each of the components represents a distributed application and they communicate with each other to carry out their tasks.

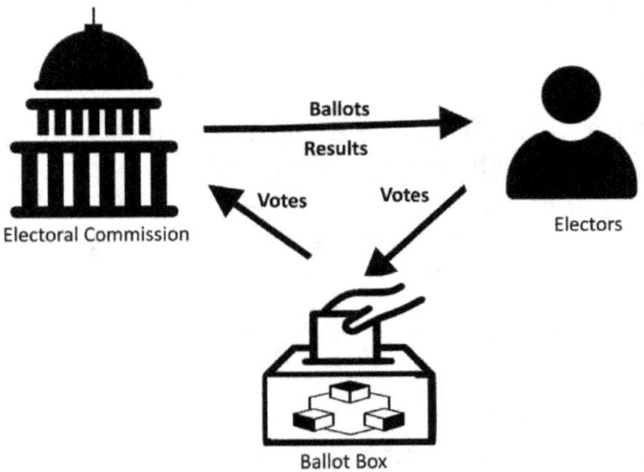

Figure 1. Electronic vote architecture.

Electoral Commission - The body responsible for overseeing the entire electoral process. It is this commission which draws up the lists of candidates which will be used to produce the ballot papers which will be presented to the voters. Once the election is over, this body is responsible for publishing the results of the election. It is also responsible for providing security and certification services for the electoral process.

Ballot Box - The entity responsible for storing the election votes. This entity receives the electors' votes, verifies their validity with the electoral commission and stores them in a secure and reliable manner. This entity serves as an interface to a service that stores the votes on a blockchain network.

Electors - Citizens entitled to vote. Voters authenticate themselves to the electoral co-mission, which returns the ballot papers to them if the authentication is successful. The voters choose their candidate by marking their preference on each of the ballot papers, or not choosing one at all, allowing a blank ballot. Once the ballot papers have been completed and certified by the Electoral Commission the votes are delivered to the Ballot box.

Electoral Commission

The class diagram (Figure 2) represents the classes that are managed by the Electoral Commission.

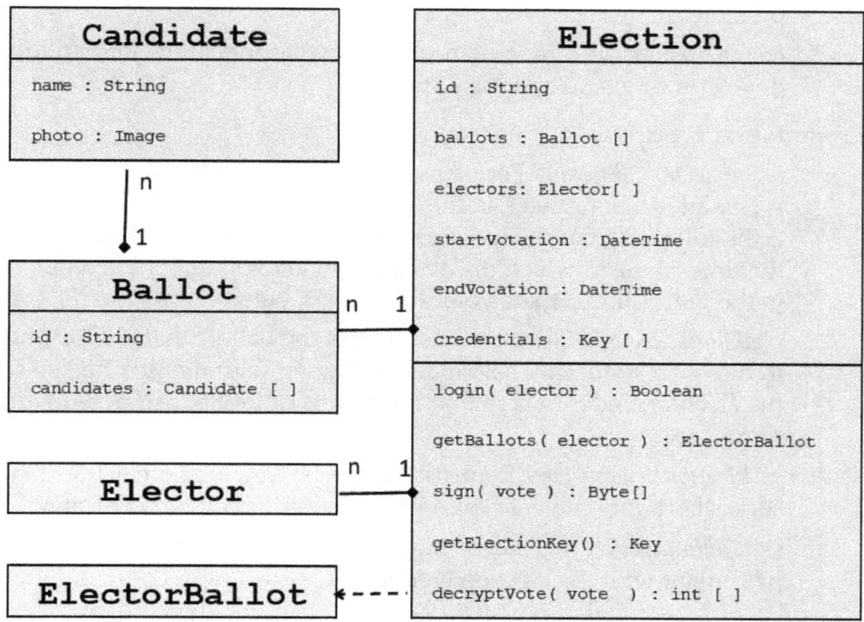

Figure 2. Class diagram of electoral commission.

- **Candidate**: Represents a candidate for election.
  Attributes:
  - **name: String** - Name of the candidate. This name can be the name of a person, of a party or of a list.
  - **photo: String** - Picture or logo of the candidate.
- **Ballot**: Represents the ballot paper in the election.
  Attributes:
  - *Id: String* - name of the ballot paper such as local council, district council or presidentials.
  - *candidates: Candidate []* - list of candidates running for the election.
- **Election**: This entity is responsible for managing the election.
  Attributes:
  - *Id: String - identifier* of the election (e.g., Election 2022, presidentials 2019).

- *Ballots: Ballot[]* – Set of paper or digital votes that are part of the electoral act.
- *Voterss: Elector[]* - A list of people who can vote in the election.
- *The start and end time of voting:* - Time period in which voting is possible.
- *credentials: Key[]* - Credentials of the election that allow encrypting, de-encrypting, and signing data.

Prompts and Processes:

- *login(elector): Boolean* - Performs authentication of the elector, which is passed as a parameter, in the election. This operation is successful if the following conditions are simultaneously met: (1) It is within the time frame in which the election can take place; (2) The voter is in the voters list; (3) The voter has not yet voted.
- *getBallots(elector): ElectorBallot* - Creates the ballots that will be sent to the voter to fill in. The ballots are signed with the private key of the election credentials so that the ballots cannot be tampered with or reused.
- *getElectionKey(): Key* - Returns the public key of the election that serves both for data encryption and digital signature verification.
- *sign(data): Byte[]* - Allows signing the data that is passed as parameter with the election credentials.
- *decrypt(vote): int[]* - Decrypts the voter's choices. This method can only be executed when the voting is over.

Electors

The class diagram in Figure 3 represents the classes that are managed by the voter application.

- **Elector**: Represents a voter.

  Attributes:

  - *id: String* – Identifier of the elector
  - *name: String* – Name of the elector
  - *photo: Image* – Photo of the elector
  - *credentials: Key []* – User credentials that allow encrypting and decrypting data,

- **ElectorBallot**: Represents the ballot paper delivered to the voter. The ballot is created specifically for the voter with one or more ballot papers, and is digitally signed to ensure its authenticity. This

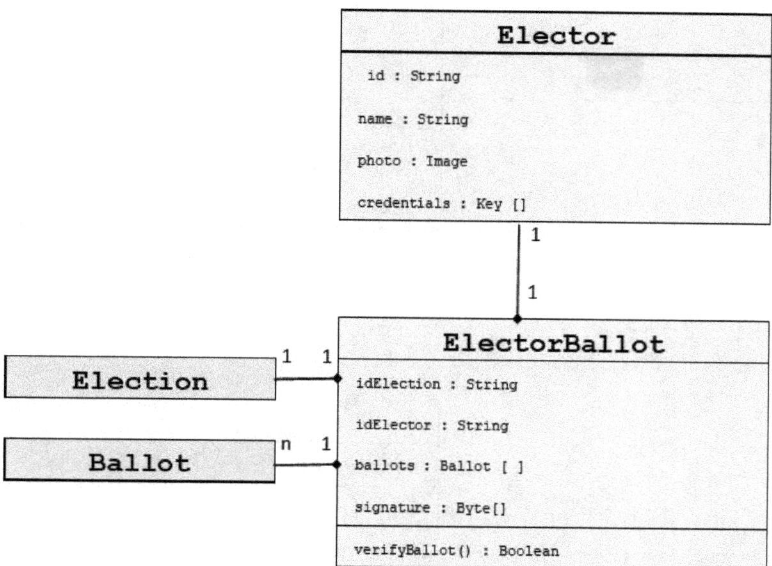

Figure 3. Class diagram of electors.

personalization and signature ensure that fraudulent copies of ballot papers cannot be made.

Attributes:

- *idElection*: *String* - Identifier of the election
- *idElector: String* - Identifier of the voter
- *ballots: Ballot[]* - Ballot paper list
- *signature* - Digital signature of the Electoral Commission to ensure its authenticity.

Methods:

- *overifyBallot(): Boolean*: checks whether the ballot paper is valid by using the digital signature.

BallotBox

The class diagram in Figure 4 represents the classes that are managed by the ballot box application.

- **Vote**: Represents completed ballot papers. This entity must guarantee the anonymity of the voter, the confidentiality of the choices and ensure that the ballot paper is valid.

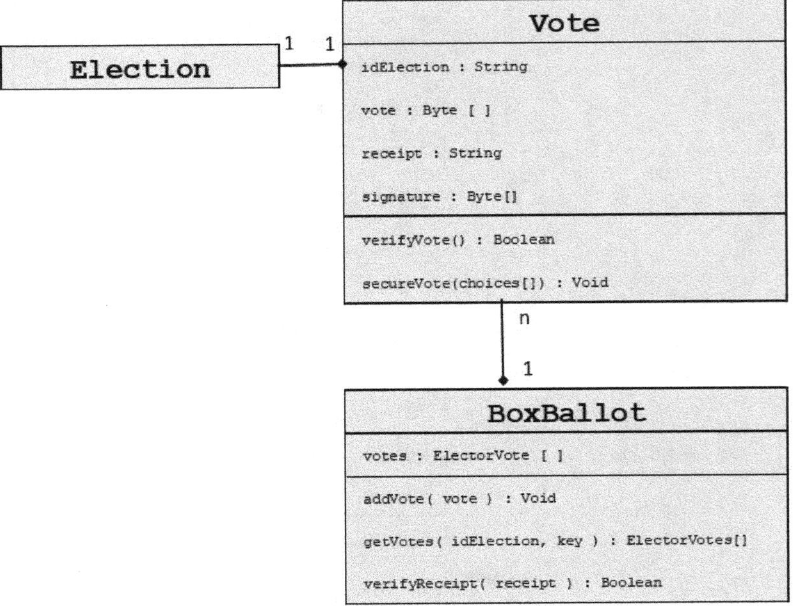

Figure 4. Class diagram of BallotBox.

Attributes:

- *idElection: String* - Identifier of the election to which the vote belongs.
- *vote: Byte[]* - Choices made by the user in the election. These choices must remain anonymous and therefore their content is encrypted.
- *receipt: String* - Receipt given to the voter confirming that he/she has exercised his/her right to vote and serves to verify that the vote has been registered.
- *signature: Byte[]* - Digital signature of the Electoral Commission that guarantees the integrity and authenticity of the vote.

Methods:

- *secureVote(choices[]) : Byte[]* - This makes the user's choices secure by generating an auditable, integer and confidential vote through the automatic completion of its attributes. This process guarantees that the *choices[]* parameter is stored securely and that none of the entities involved in its security know its content.

  The vote attribute is generated by encrypting the voter's choices, *choices[]*, with the election key. Due to the fact that the encryption key is public and the voter's choice options are limited, confidentiality can only be ensured if a sufficiently large random data set is added to the user's choices. Without the addition of these data the decryption process is trivial.

The *receipt* attribute is generated with a set of random data which, as it has no correspondence with the voter in order to ensure the anonymity of the vote. The usefulness of this attribute is related to the verification, by the voter, if his vote is present in the ballot box.

The *signature* attribute represents the signature of the vote made by the Electoral Commission. This signature is done blindly, which means that the entity that signs has no knowledge about the signed content.

- *verifyVote()* - Checks if the vote is valid through the digital signature of the Electoral Commission.

4.3 System Implementation

Figure 5 shows the electronic voting application. This application is only accessible after the user is successfully authenticated (*login*). The figure shows two ballot papers, Local council and District council, and the candidates running for Local Council (Candidate 1, Candidate 2,

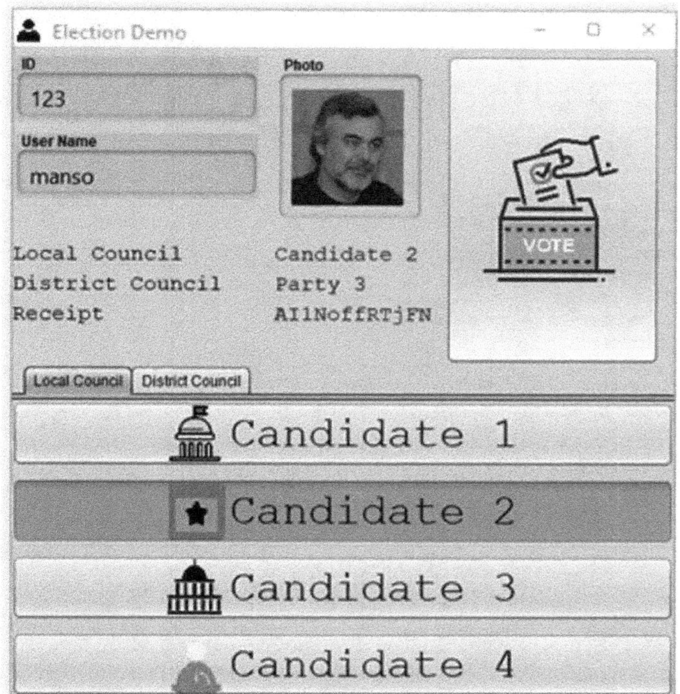

Figure 5. Electors application.

Candidate 3 and Candidate 4). These ballot papers were provided by the getBallots method. In the application the voter has already made his choices (Candidate 2 and Party 3).

The ballots were generated by the Electoral Commission and their transmission was encrypted by the voter's credentials so that the ballots can only be used once. When the application receives the encrypted ballots, it decrypts them and verifies the Electoral Commission signature to ensure their authenticity.

At this point, the system has generated a voting receipt: AI1NoffRTjFN.

When the voter presses the Vote button, the *secureVote* method is called, which implements the anonymity, confidentiality and authenticity of the voter's choices to submit them to the BallotBox application and delivers the voting receipt (AI1NoffRTjFN) to the voter.

The BallotBox application, Figure 6, is a node in a network of computers that support a blockchain. This application has a server that provides services to the other components of the system. Votes are received from the voter application; after their validity and integrity are verified with the Electoral Commission (*verifyVote*), they are requested to be included in a block of the blockchain (*addVote*). Once the block is validated for the network it becomes available for consultation through the receipt that was

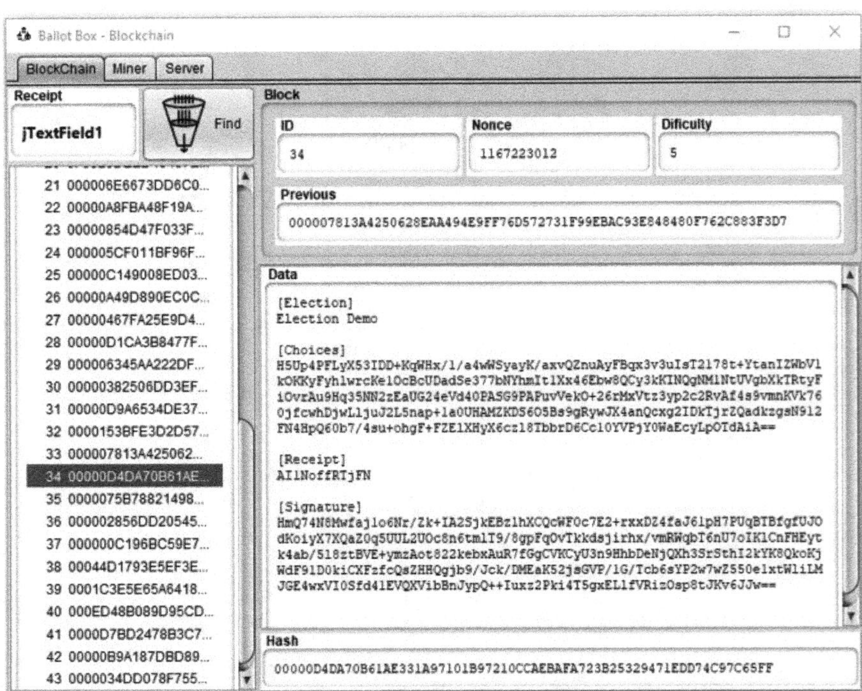

Figure 6. Ballot Box application.

provided to the voters when they voted. This way, the voters know that their vote will count in the election.

Figure 6 shows the vote that the voter in Figure 5 cast through the AI1NoffRTjFN receipt. This receipt is known only to the voter and therefore only he knows that this record represents his vote. The voter's vote, [*choices*], is encrypted with the election key and is therefore confidential. Although the voter's choice has only two options, a large amount of random data is added to it in order to guarantee confidentiality.

Figure 7 shows the application that represents the Electoral Commission, which provides various services to the other components of the system through a server that runs in the background. The application manages the electors who have the right to vote, manages the electoral lists and displays the election results.

In the image in Figure 7 we can see the votes that the voters submitted during the election process. The votes were decrypted by the *decryptVote* method and the election results can be computed. In this application the voters can check if their vote counted towards the result.

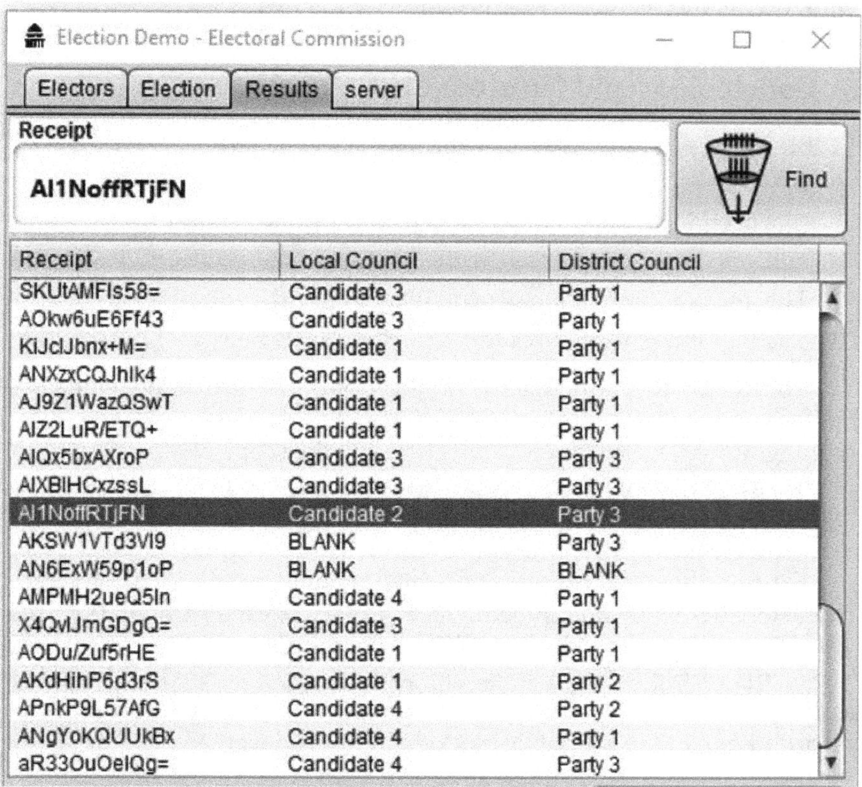

Receipt	Local Council	District Council
SKUtAMFIs58=	Candidate 3	Party 1
AOkw6uE6Ff43	Candidate 3	Party 1
KiJclJbnx+M=	Candidate 1	Party 1
ANXzxCQJhIk4	Candidate 1	Party 1
AJ9Z1WazQSwT	Candidate 1	Party 1
AIZ2LuR/ETQ+	Candidate 1	Party 1
AIQx5bxAXroP	Candidate 3	Party 3
AIXBIHCxzssL	Candidate 3	Party 3
AI1NoffRTjFN	Candidate 2	Party 3
AKSW1VTd3VI9	BLANK	Party 3
AN6ExW59p1oP	BLANK	BLANK
AMPMH2ueQ5In	Candidate 4	Party 1
X4QvlJmGDgQ=	Candidate 3	Party 1
AODu/Zuf5rHE	Candidate 1	Party 1
AKdHihP6d3rS	Candidate 1	Party 2
APnkP9L57AfG	Candidate 4	Party 2
ANgYoKQUUkBx	Candidate 4	Party 1
aR33OuOelQg=	Candidate 4	Party 3

Figure 7. Electoral Commission Application.

4.4 System Testing

The final system was tested to choose the day on which the class would present the projects. The teacher provided the days that could be chosen and all students were able to vote. The election was set up and the voting deadline set in one week. A private blockchain was created to securely store votes, and the application was made available to all students who could vote.

During voting time, eligible students were able to exercise their right to vote, prove that their vote was recorded through the ticket, and were prevented from voting more than once. It was also possible to verify the confidentiality of the vote, encrypted by the public key of the election.

After the voting deadline, the Electoral Commission decrypted the data present in the BallotBox and published them in decrypted form. The Electoral Commission also made the private key of the election available so that voters could decrypt it and prove that their choices are the ones that were published. The anonymity of voters' votes, even after publication of the decryption key, is maintained as there is no relationship between the vote receipt and the voter who produced it.

5. Conclusions and Future Work

In this paper we present a distributed secure electronic voting system with 3 components:

1. The Electoral Commission that manages the parameters of the election and provides the voter authentication and vote certification services;
2. The application that allows voters to vote and
3. The Ballot Box that stores the votes in a blockchain.

To ensure the authenticity of the votes we use blind signatures and to ensure the confidentiality of the vote we use asymmetric key cryptography. The system was developed in the scope of a curricular unit of Distributed Systems, as part of the Degree in Informatics Engineering of the Polytechnic Institute of Tomar and was tested in the laboratory with a private blockchain.

During the tests it was possible to confirm that all the initial characteristics were guaranteed. The eligibility of each student, the single vote, data encryption, the possibility to check if the vote is unchanged. These are very positive results, but wider testing is needed. Therefore, as future work it is necessary to make scalability tests for its use in real environment and use a public blockchain to support the votes.

Portugal still needs to revise its legislation so that the possibility of electronic voting becomes feasible (Braz, 2021). Through this proof of concept it is possible to understand how an electronic election system based on blockchain could be implemented. Its benefits are obvious; the first is economic, due to the reduction in paper costs and people; the second is environmental, due to the reduction in waste and travel; and the third is universal access, as any Portuguese person eligible to vote will be able to do so from anywhere. Security issues will have to be tested, and a comprehensive study will have to be conducted as blockchain technology also has some drawbacks (Jafar et al., 2021; Park et al., 2021).

Finally, from the point of view of the curricular exercise, involving students in real issues, studying and testing solutions, this created great motivation and engagement. Using this type of initiative brings a double benefit: facilitating the development of a solution to a real problem, while at the same time training students for the real context of their future career. Initially, this challenge presented itself as impossible for the students, and therefore was demotivating. Overcoming this with collaborative work, motivated students, empowering them for future employment scenarios.

References

Alharby, M. and Moorsel, A. Van. (2017) Blockchain-based smart contracts: A systematic mapping study. Academy Industry Research Collaboration Center (AIRCC). https://doi.org/10.5121/csit.2017.71011.

Braz, B. (2021). A proposal for the use of blockchain in the Portuguese voting system. [Master's thesis, Universidade Nova de Lisboa]. http://hdl.handle.net/10362/116180.

Chaum, D. (1983). Blind signatures for untraceable payments. pp. 199–203. *In*: Chaum, D., Rivest, R.L. and Sherman, A.T. (eds.). Advances in Cryptology. Springer: Boston.

Huang, J., Obaidat, M.S., Vijayakumar, P., Luo, M. and Choo, K.-K. R. (2021). The application of the blockchain technology in voting systems: A review. ACM Comput. Surv., 54(3): 1–28. https://doi.org/10.1145/3439725.

Jafar, U., Aziz, M.J.A. and Shukur, Z. (2021). Blockchain for electronic voting system—review and open research challenges. Sensors, 21(17): 5874. MDPI AG. http://dx.doi.org/10.3390/s21175874.

Maesa, F. and Mori, P. (2020). Blockchain 3.0 applications survey. Journal of Parallel and Distributed Computing, 138: 99–114. https://doi.org/10.1016/j.jpdc.2019.12.019.

Monteiro, J. (2019). Blockchain-based Decentralized Application for Electronic Voting using an Electronic ID. [Master's thesis, Universidade da Beira Interior]. http://hdl.handle.net/10400.6/9230.

Nakamoto, S. (2008). Bitcoin: A peer-to-peer electronic cash system, Decentralized Business Review, Availabul in https://bitcoin.org/bitcoin.pdf.

Natarajan, H., Krause, S. and Gradstein, H. (2017). Distributed ledger technology and blockchain. FinTech Note, 1. World Bank, Washington, DC. http://hdl.handle.net/10986/29053.

Park, S., Specter, M., Narula, N. and Rivest, R. (2021). Going from bad to worse: From internet voting to blockchain voting. Journal of Cybersecurity, 7(1): 1–15. https://doi.org/10.1093/cybsec/tyaa025.

Russell, M. and Zamfir, I. (2018). Digital Technology in Elections - Efficiency versus Credibility? European Parliamentary Research Service. Accessed December 2019. http://www.europarl.europa.eu/RegData/etudes/BRIE/2018/625178/EPRS_BRI(2018)625178_EN.pdf.

Zhang, C., Wu, C. and Wang, X. (2020). Overview of blockchain consensus mechanism. In Proceedings of the 2020 2nd International Conference on Big Data Engineering, New York, NY, USA, pp. 7–12. https://doi.org/10.1145/3404512.3404522.

Chapter 12

The Relationship Between Blockchain Technology and Sustainability
A Bibliometric Analysis

Soraya González-Mendes,[1,*] *Fernando García-Muiña*,[1] *Sara Alonso-Muñoz*,[1] *Carlos J. Costa*[2] and *Rocío González-Sánchez*[1]

1. Introduction

In a context in which technology is the order of the day, blockchain is an emerging technology that stands out for its decentralisation, transparency, traceability (Pesqueira et al., 2023), immutability, Smart Contracts, and privacy. Which shows the sustainability of supply chains improving the management of companies (Mukherjee et al., 2021; Kouhizadeh et al., 2021; Bai et al., 2021). It has the potential to change Sustainable Business Models (SBMs) significantly (Tiscini et al., 2020) promoting financial and social sustainability (Massaro et al., 2020). Blockchain is a disruptive and innovative technology that improves inter-organisational processes

[1] Business Administration (ADO), Applied Economics II and Fundaments of Economic Analysis Rey Juan Carlos University. Madrid, Spain.
[2] Advance/ISEG, Lisbon School of Economic & Management. Universidade de Lisboa.
Emails: fernando.muina@urjc.es; sara.alonso@urjc.es; cjcosta@iseg.ulisboa.pt; rocio.gonzalez@urjc.es
* Corresponding author: soraya.gonzalez@urjc.es

(Bernardino et al., 2022). It stands out for having a transparent registry, with a shared and decentralised database (Abreu et al., 2018; Saini et al., 2022; Machado et al., 2020) where the copies of immutable and encrypted information are stored in each node of the network (Mercuri et al., 2021). In the future, the development of a less energy-intensive alternative to record data would make blockchain the ideal technology to support environmental, economic, and social sustainability (Rana et al., 2019).

The development of new technologies such as blockchain play a fundamental role in supporting sustainability (Mangla et al., 2022), improving aspects of circular economy activities (CE) (Rejeb et al., 2022b; Zhang et al., 2020). Blockchain has the potential to boost circular economy practices by improving the financial, environmental, and organisational performances (Khan et al., 2021) of companies. This article aims to show blockchain technology and its relationship with sustainability, highlighting the evolution of publications, leading researchers, and topics that are treated together. It presents a review of the state of the art, proposing future lines of research since more research in this field is necessary. Particularly, this research addresses these interesting research questions (RQs): What is the evolution of this topic in the scientific literature? (RQ1) Which authors, institutions, and journals published the most on this research topic? (RQ2) Which are the most studied topics, and which are the most cited articles? (RQ3).

Consequently, this work shows the following composition. Section 2 indicates the data collection and the methods applied in the search. Section 3, presents an analysis of the results obtained in relation to the authors, countries, organisations, and journals with more publications on concepts of blockchain and sustainability. Section 4 explains the most relevant topics and the most cited articles. Finally, Section 5 presents our conclusions about the analysis, and proposes future lines of research.

2. Proposed Methodology

This bibliometric evaluates the state of the art about blockchain technologies and sustainability. This technique offers a quantitative study of the interconnections between these concepts in the scientific literature. To show research output over a given period and about this specific field (Diem and Wolter 2013). Our paper uses the VOSviewer to analyse bibliometric maps. It reveals that the larger the circle on the map, the more relevant the article (Van Eck and Waltman 2010).

Figure 1 illustrates the phases analysed through the bibliometric analysis of blockchain and sustainability concepts. The paper is divided up into two sections: (1) data source and (2) bibliometric analysis. In

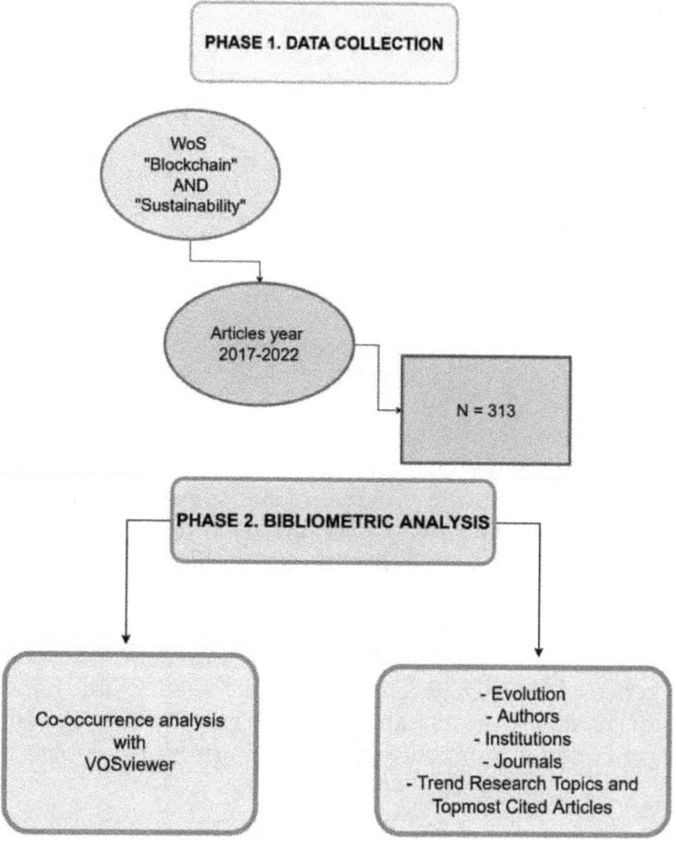

Figure 1. Proposed research methodology.

phase 1, it performs a keyword search of 'blockchain' and 'sustainability' terms in the Web of Science database, over a period between 01/01/2017 to 27/06/2022. It filters out lectures and books, considering only articles from SCI-EXPANDED, SSCI, and A&HCI. In total, 313 items were obtained, related to concepts of blockchain and sustainability. In phase 2, it performs a bibliometric analysis in VOSviewer program to understand the co-occurrence represented by nodes. It considers the principal investigators, institutions, and journals with a higher number of publications.

3. Results

3.1 Analysis of Historical Publication

Research related to blockchain technology and sustainability during the period between 2017 and 2022 has increased. A large increase in

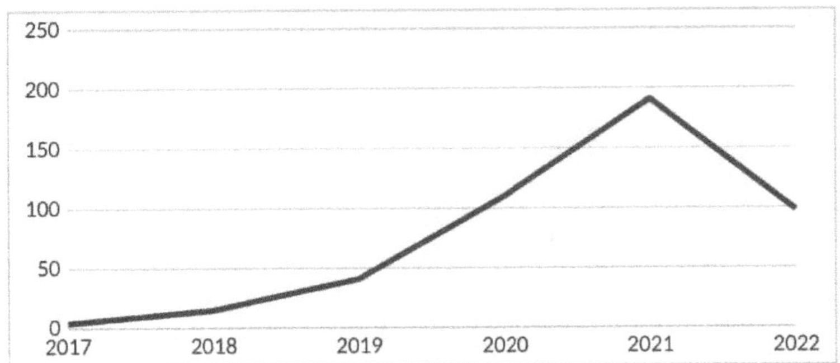

Figure 2. Evolution of the publications related to blockchain and sustainability between 2017 and June of 2022.

publications is seen in 2020 and 2021. In recent years, concerns about the increase in CO_2 emissions taking place in cities, climate change, the Covid-19 pandemic, and the progress of digitalisation make it necessary for emerging technologies to be leveraged in supporting environmental, economic, and social sustainability in society. Blockchain is one of these emerging technologies.

Figure 2 shows that the year in which the greatest interest and the most significant number of publications by researchers occurred was after the emergence of COVID-19. This reflects the need to study more in the field of blockchain and sustainability in order to mitigate challenges, such as losses generated in companies as well as improving people's quality of life.

3.2 *Authors and Institutions*

The analysis carried out on the authors explains that the largest number of authors belong to Universities in the United States, the United Kingdom, and China. It shows the topmost productive authors highlighting institutions such as Worcester Polytechnic Institute, or the University of Rhode Island, as seen in Table 1. The results show that the authors with the highest number of published papers are Joseph Sarkis (with 14 papers and 1,492 citations), Mahtab Kouhizadeh (6 documents and 1,223 citations), and Sachin Kumar Mangla (5 documents and 35 citations). Most of these authors' articles focus on improving aspects of supply chains while remaining more sustainable.

Kouhizadeh and Sarkis (2018) consider blockchain an innovative technology that can enhance sustainable supply chains, especially ecological ones. Kumar et al. (2020) propose the creation of the PRODCHAIN platform by introducing the qualifications-based consensus mechanism

Table 1. Authors with the largest number of documents.

R	Authors	Institution	Country	D	C
1.	Sarkis, Joseph	Worcester Polytechnic Institute	United States	14	1,492
2.	Kouhizadeh, Mahtab	University of Rhode Island	United States	6	1,223
3.	Mangla, Sachin Kumar	University of Plymouth	United Kingdom	5	35
4.	Yu, Zhang	Chang′an University	China	5	92
5.	Bai, Chunguang	University of Electronic Science and Technology of China	China	4	150
6.	Kang, Kai	University of Hong Kong	China	4	220
7.	Zhang, Abraham	University of Stirling Management School	United Kingdom	4	270
8.	Zhong, Ray Y.	University of Hong Kong	China	4	208
9.	Dora, Manoj	Anglia Ruskin University	United Kingdom	3	13
10.	Gunasekaran, Angappa	California State University, Bakersfield	United Kingdom	3	201

Abbreviations: R = rank; D = number of documents; C = number of citations.

called Proof of Accomplishment (PoA) to monitoring products, enhancing social and financial sustainability. Yu et al. (2022) observe that blockchain is a digital accounting technology, distributed, safe, and visible, with traceability and transparency to alleviate environmental and supply chain problems. Bai and Sarkis (2020) state that blockchain can support supply chain transparency and, to incorporate technical and transparency attributes of the sustainable supply chain, introduce a new method for evaluating and selecting blockchain technology.

3.3 *Analysis of Journals*

Table 2 indicates the ten most productive journals, which it considered from WoS Database, Q1 and Q2 JCR, and SCOPUS. To recognise the main journals, it relies on the number of papers published and the number of citations. The three journals with the most documents are *Sustainability, Journal of Cleaner Production,* and *Business Strategy and the Environment*. The journals with the most citations are the *International Journal of Information Management* and the *International Journal of Production Research*. All journals are related to technology and sustainability.

Table 2. Topmost productive journals.

R	Source	D	C
1.	Sustainability	66	835
2.	Journal of Cleaner Production	18	347
3.	Business Strategy and the Environment	10	53
4.	International Journal of Production Research	8	938
5.	Annals of Operations Research	8	50
6.	Energies	8	29
7.	International Journal of Information Management	6	1,148
8.	IEEE Access	6	87
9.	Energy Research & Social Science	6	84
10.	Technological Forecasting and Social Change	6	28

Abbreviations: R = rank; D = number of documents; C = number of citations.

3.4 Trend Research Topics and Topmost Cited Articles

3.5 Relevant Research Topics

In Figure 3, the analysis by occurrences of the keywords performed by the VOSviewer software can be seen. It indicates the most searched topics related to Blockchain and Sustainability. From the 313 articles examined, it recovered 1561 keywords. It selected keywords with a minimum of 15 occurrences, G and 40 keywords met this criterion. It shows four lines of research on the relationship between blockchain and sustainability.

The red cluster contains the keyword 'blockchain' related to 'supply chains', 'industry 4.0' or 'performance' which is linked with 'sustainability' and other keywords such as 'management' and 'technologies'. Technological advances such as blockchain and other emerging technologies show more efficiency and sustainability in supply chains (Rejeb et al., 2022a). The yellow node is associated to 'sustainability' and 'technologies' in relation to 'challenges' and 'management'. This cluster highlights the relationship between 'transparency' and 'information'. Emerging technologies such as blockchain allow for changes in the management of companies, showing transparency in supply chains (Sahoo et al., 2022). The blue node of keywords represents the 'future' and the 'security' of 'smart contracts', 'big data', 'digitalisation' and 'internet of things'. Blockchain is a technology emerging from Industry 4.0 that together with others such as Big Data or Artificial Intelligence has the potential to achieve sustainability (Nabeeh et al., 2022). The

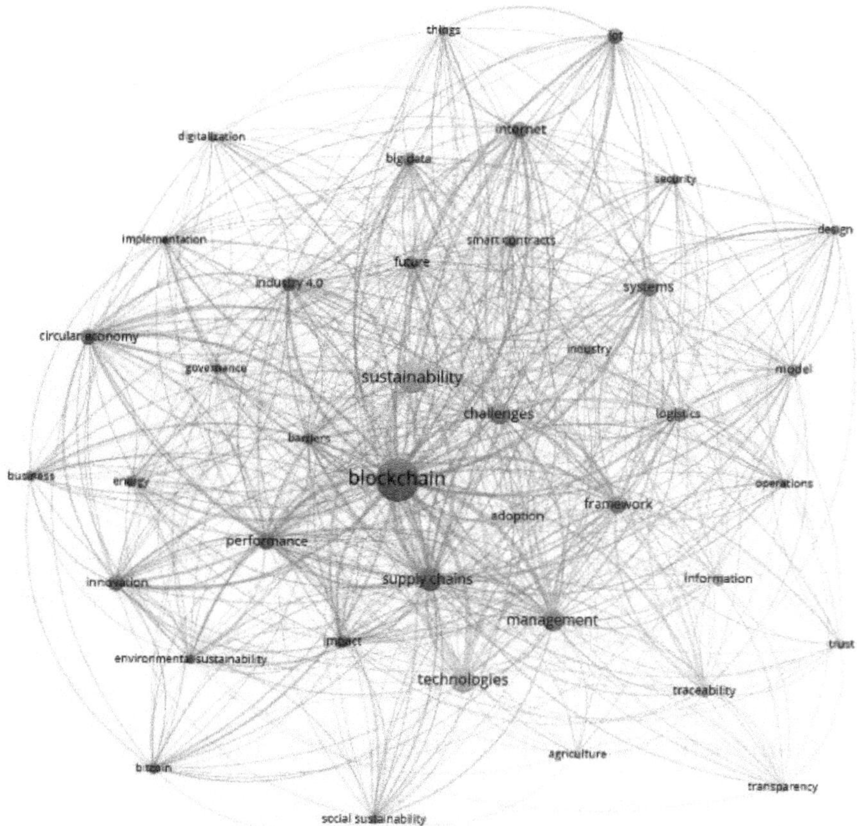

Figure 3. Co-occurrence analysis.

green cluster links 'challenges and 'management' with 'climate change' with 'sustainability' and 'technologies'. In the context of climate change, technologies such as blockchain allow society to be sustainable with the environment when using consensus mechanisms like Proof of Stake (PoS)-validators with the most quantity of cryptocurrencies validate transactions- in comparison with the use of Proof of Work (PoW) (e.g., Bitcoin)-miners confirms transactions using a lot of quantity of energy to solve complex mathematical functions- (Dorfleitner et al., 2021).

As seen from Table 3, the hottest topic was blockchain (259) followed by sustainability (144), technologies (89), supply chains (79), and management (70). Blockchain has the potential to contribute to sustainability and to clear barriers to the Circular Economy (CE) (Erol et al. 2022), improving supply chain management (Di Vaio and Barriale (2020).

Table 3. Top ten keywords of blockchain research based on the occurrences.

Rank	Keyword	Occurrences	Total Link Strength
1.	Blockchain	259	1,056
2.	Sustainability	144	705
3.	Technologies	89	461
4.	Supply Chains	79	471
5.	Management	70	377
6.	Challenges	67	383
7.	Framework	52	312
8.	Systems	50	266
9.	Performance	41	232
10.	Internet	40	226

3.6 *Topmost Cited Articles*

As seen in Table 4, the ten most cited articles are related to the potential that blockchain technology has in supply chains, the creation of Smart Cities (cities supporting sustainability), and the application of other technologies such as AI or Machine Learning.

4. Conclusions and Future Lines of Research

In a challenging environment with problems such as climate change, blockchain is an emerging decentralised, traceable, and immutable technology that improves the management of companies (Hong et al., 2021). In terms of sustainability, although blockchain may be sustainable when using consensus mechanisms such as PoS, when PoW is used a huge amount of energy is generated (Zheng et al., 2017). Thus, there are still some challenges to be addressed towards more sustainable blockchain application. Our paper answers the research questions related to blockchain and sustainability, analysing its evolution (RQ1), authors, countries, institutions, and journals (RQ2), and the relevant research issues, conducting a co-occurrence analysis (RQ3). With the outbreak of COVID-19 in 2019, there was a greater number of publications, probably due to companies' need to improve their situation by betting on emerging technologies such as blockchain and seeking sustainability (RQ1). Most of the authors belong to universities in the United States, China, and the United Kingdom, and the journals with the most documents and citations are *Sustainability* and *International Journal of Information Management*, respectively (RQ2). The research shows the topics most related to

Table 4. Topmost cited articles with contributions and limitations.

R	Authors	Aspect	Contributions	Limitations	C
1.	Saberi et al. (2019)	Strategic	Blockchain allows accounting for data permits, transparency, security, and traceability.	The globalisation of supply chains makes it difficult to manage and control them.	706
2.	Kshetri (2018)	Economic	Blockchain increases supply chain transparency and accountability.	Challenges blockchain has to overcome: global supply chain regulations, and not all countries are ready to adopt blockchain.	521
3.	Hughes et al. (2019)	Environment	Blockchain has the potential to contribute to the UN Sustainable Development Goals.	A dearth of studies related to the significant focus of Blockchain within the literature on Information Systems (IS) and Information Management (IM).	202
4.	Allam and Dhunny (2019)	Environment and culture	Cities increasingly integrate technologies that support sustainability, favouring the creation of Smart Cities.	New technologies can help solve problems of confidentiality and ethics but there is a scarcity of legislation.	200
5.	Kamble et al. (2020)	Strategic	Blockchain has improved the gist of supply chains by eliminating trust issues.	More studies related to supply chains are needed in other countries such as developed countries.	190
6.	Cole et al. (2019)	Strategic	Blockchain enhances product safety and security improving supply chain management	Developing more studies about improving the sustainability in the supply chain is essential.	163
7.	Kouhizadeh and Sarkis (2018)	Environment	The green supply chain management could benefit from applying blockchain.	There is a scarcity of studies about the sustainability supply chain applying blockchain.	160

Table 4 contd. ...

...Table 4 contd.

R	Authors	Aspect	Contributions	Limitations	C
8.	Kouhizadeh et al. (2021)	Environment	Blockchain improves the efficiency, transparency, and traceability of supply chain management.	It could be development more research about the barriers to adopted blockchain..	159
9.	Hastig and Sodhi (2020)	Strategic	It is necessary to develop traceability systems in the supply chain for improving sustainability performance enhancing supply chain coordination.	More empirical studies could be done.	141
10	Bai et al. (2020)	Strategic	Analytical models are examined to understand and analyse the decisions of blockchain applications in supply chains.	It could be realised in other analyses of the analytical models in the future.	136

Abbreviations: R = rank; C = number of citations.

blockchain and sustainability. Among them are the improvement of efficiency and transparency in the supply chain using blockchain, the enhancement of sustainability using emergent technologies such as Big Data and AI with blockchain, as well as the challenges in the consensus mechanism to improve sustainability (RQ3).

This article adds knowledge about blockchain and sustainability identifying the main contributions: strategic, cultural, technological, and environmental factors. This article aims to create more investigations in this field related to the application of new technologies that allow the development of sustainable practices related to the environment. The results shown can serve future researchers in the use of emerging technologies that allow supporting sustainability by creating Smart Cities.

Regarding limitations, it must be noted only WoS articles have been considered. Bibliometric analyses could be performed with SCOPUS and other programs such as SciMat. A systematic literature review and a new bibliometric study will help continue the research on this subject. A research agenda could be created for future research. This investigation provides data and useful information for further research. Answering the research questions in this chapter could help other researchers or practitioners to continue studying blockchain and sustainability and understand the 'state-of-the-art' according to the collected publications up to June 2022. This paper considers the value of implementing blockchain technology with other emergent technologies like Big Data or Artificial Intelligence in companies to include sustainability in the management sector.

References

Abreu, P.W., Aparicio, M. and Costa, C.J. (2018). Blockchain technology in the auditing environment. 2018 13th Iber. Conf. Inf. Syst. Technol. (CISTI), Cáceres, Spain, 1–6.

Allam, Z. and Dhunny, Z.A. (2019). On big data, artificial intelligence, and smart cities. Cities, 89: 80–91.

Bai, C. and Sarkis, J. (2020). A supply chain transparency and sustainability technology appraisal model for blockchain technology. Int. J. Prod. Res., 58(7): 2142–2162.

Bai, C., Zhu, Q. and Sarkis, J. (2021). Joint blockchain service vendor-platform selection using social network relationships: A multi-provider multi-user decision perspective. Int. J. Prod. Econ., 238: 108165.

Bernardino, C., Costa, C.J. and Aparicio, M. (2022). Digital Evolution: Blockchain field research. 2022 17th Iber. Conf. Inf. Syst. Technol. (CISTI), Madrid, Spain, 1–6.

Cole, R., Stevenson, M. and Aitken, J. (2019). Blockchain technology: Implications for operations and supply chain management. J. Supply Chain Manag., 24(4): 469–483.

Diem, A. and Wolter, S. (2013). The use of bibliometrics to measure research performance in education sciences. Res. High. Education, 54(1): 86–114.

Di Vaio, A. and Varriale, L. (2020). Blockchain technology in supply chain management for sustainable performance: Evidence from the airport industry. Int. J. Inf. Manage., 52: 102014.

Dorfleitner, G., Muck, F. and Scheckenbach, I. (2021). Blockchain applications for climate protection: A global empirical investigation. Renew. Sust. Energ., 149: 111378.

Erol, I., Murat, A.I., Peker, I. and Searcy, C. (2022). Alleviating the impact of the barriers to circular economy adoption through blockchain: An investigation using an integrated MCDM-based QFD with hesitant fuzzy linguistic term sets. Comput. Ind. Eng., 165: 107962.

Hastig, G.M. and Sodhi, M.S. (2020). Blockchain for supply chain traceability: Business requirements and critical success factors. Prod. Oper. Manag., 29: 935–954.

Hong, W., Mao, J., Wu, L. and Pu, X. (2021). Public cognition of the application of blockchain in food safety management Data from China's zhihu platform. J. Clean. Prod., 303: 127044.

Hughes, L., Dwivedi, Y.K., Misra, S.K., Rana, N.P., Raghavan, V. and Akella, V. (2019). Blockchain research, practice, and policy: Applications, benefits, limitations, emerging research themes and research agenda. Int. J. Inf. Manage., 49: 114–129.

Kamble, S.S., Gunasekaran, A. and Sharma, R. (2020). Modelling the blockchain enabled traceability in agriculture supply chain. Int. J. Inf. Manage., 52: 101967.

Khan, S.A.R., Razzaq, A., Yu, Z. and Miller, S. (2021). Industry 4.0 and circular economy practices: A new era business strategies for environmental sustainability. Bus. Strategy. Environ., 30(8): 4001– 4014.

Kouhizadeh, M., Saberi, S. and Sarkis, J. (2021). Blockchain technology and the sustainable supply chain: Theoretically exploring adoption barriers. Int. J. Prod. Econ., 231: 107831.

Kouhizadeh, M. and Sarkis, J. (2018). Blockchain practices, potentials, and perspectives in greening supply chains. Sustainability, 10: 3652.

Kshetri, N. (2018). 1 blockchain's roles in meeting key supply chain management objectives. Int. J. Inf. Manage., 39: 80–89.

Kumar, G., Saha, R., Buchanan, W.J., Geetha, G., Thomas, R., Rai, M.K. et al. (2020). Decentralised accessibility of e-commerce products through blockchain technology. Sustain. Cities. Soc., 62: 102361.

Machado, A., Sousa, M.J. and Rocha, Á. (2020). Blockchain technology in education. In 2020 The 4th International Conference on E-commerce, E-Business and E-Government (ICEEG 2020), June 17–19, 2020, Arenthon, France. ACM, New York, NY, USA, 5 pages.

Mangla, S.K., Kazançoğlu, Y., Yıldızbaşı, A., Öztürk, C. and Çalık, A. (2022). A conceptual framework for blockchain-based sustainable supply chain and evaluating implementation barriers: A case of the tea supply chain. Bus. Strategy. Environ., 1– 24.

Massaro, M., Dal Mas, F., Chiappetta Jabbour, C.J. and Bagnoli, C. (2020). Crypto-economy and new sustainable business models: Reflections and projections using a case study analysis. Corp. Soc. Responsib. Environ. Manag., 27: 2150–2160.

Mercuri, F., Gaetano, C. and Federica, R. (2021). Blockchain technology and sustainable business models: A case study of devoleum. Sustainability, 10. 10: 5619.

Mukherjee, A.A., Singh, R.K., Mishra, R. and Bag, S. (2022). Application of blockchain technology for sustainability development in agricultural supply chain: Justification framework. Oper. Manag. Res., 15: 46–61.

Nabeeh, N.A., Abdel-Basset, M., Gamal, A. and Chang, V. (2022). Evaluation of production of digital twins based on blockchain technology. Electronics, 11(8): 1268.

Pesqueira, A., Sousa, M.J. and Bolog, S. (2023). Implementation of big data and blockchain for health data management in patient health records. *In*: Anwar, S., Ullah, A., Rocha, Á. and Sousa, M.J. (eds.). Proceedings of International Conference on Information Technology and Applications. Lec. Not. Net. Syst, 614. Springer, Singapore.

Rana, R.L., Giungato, P., Tarabella, A. and Tricase, C. (2019). Blockchain applications and sustainability issues. Amfiteatru Econ., 21(3): 861–870.

Rejeb, A., Rejeb, K., Abdollahi, A., Zailani, S., Iranmanesh, M. and Ghobakhloo, M. (2022a). Digitalisation in food supply chains: A bibliometric review and key-route main path analysis. Sustainability, 14(1): 83.

Rejeb, A., Rejeb, K., Keogh, J.G. and Zailani, S. (2022b). Barriers to blockchain adoption in the circular economy: A fuzzy delphi and best-worst approach. Sustainability, 14(6): 3611.

Saberi, S., Kouhizadeh, M., Sarkis, J. and Shen, L. (2019). Blockchain technology and its relationships to sustainable supply chain management. Int. J. Prod. Res., 57(7): 2117–2135.

Sahoo, S., Kumar, S., Sivarajah, U., Lim, W.M., Westland, J.C. and Kumar. 2022. Blockchain for sustainable supply chain management: Trends and ways forward. Electron. Commer. Res.

Saini, H., Dash, S., Pani, S., Sousa, M. and Rocha, Á. (2022). Blockchain-based raw material shipping with PoC in Hyperledger Composer. Comput. Sci. Inf., 19(3): 1075–1092.

Tiscini, R., Testarmata, S., Ciaburri, M. and Ferrari, E. (2020). The blockchain as a sustainable business model innovation. Manag. Decis., 58(8): 1621–1642.

Van Eck, N.J. and Waltman, L. (2010). Software survey: VOSviewer, a computer program for bibliometric mapping. Scientometrics, 84(2): 523–538.

Vogel, R. and Güttel, W.H. (2013). The dynamic capability view in strategy management: A bibliometric review. Int. J. Manag. Rev., 15: 426–446.

Yu, Z., Umar, M. and Rehman, S.A. (2022). Adoption of technological innovation and recycling practices in automobile sector: under the Covid-19 pandemic. Oper. Manag. Res., 15: 298–306.

Zhang, A., Zhong, R.Y., Farooque, M., Kang, K. and Venkatesh,V.G. (2020). Blockchain-based life cycle assessment: An implementation framework and system architecture. Resour. Conserv. Recycl., 152: 104512.

Zheng, Z., Xie, S., Dai, H., Chen, X. and Wang, H. (2017). An overview of blockchain technology: Architecture, consensus, and future trends. In 2017 IEEE International Congress on Big Data (BigData Congress): 557–564. IEEE.

Index